Powers of Two

Edwin A. Valentijn

Editor

Powers of Two

The Information Universe — Information
as the Building Block of Everything

With a Foreword by Robbert Dijkgraaf

 Springer

Editor
Edwin A. Valentijn
Kapteyn Astronomical Institute
University of Groningen
Groningen, The Netherlands

Foreword by
Robbert Dijkgraaf
Institute for Advanced Study
Princeton, NJ, USA

ISBN 978-3-030-58347-7   ISBN 978-3-030-58345-3   (eBook)
https://doi.org/10.1007/978-3-030-58345-3

Cover photo: Walkabout Photo Guides/Shutterstock.com

This Springer imprint is published by the registered company Springer Nature Switzerland AG
The registered company address is: Gewerbestrasse 11, 6330 Cham, Switzerland

THE
POWERS
OF TWO

The Information Universe — Information as the Building Block of Everything

"Getting rid of Information is the best instrument you have"

Ivo van Hove ▎ theatre and artistic director of the International Theatre Amsterdam and many Broadway productions

at TV: VPRO Zomergasten 2019

Authored & edited by

Edwin A. Valentijn

WITH CONTRIBUTIONS BY:
PROF KATRIN AMUNTS | UNIVERSITY OF DÜSSELDORF
PROF CHRIS VAN DEN BROECK | UNIVERSITY OF AMSTERDAM
DR MARGOT BROUWER | UNIVERSITY OF GRONINGEN
PROF HEINO FALCKE | RADBOUD UNIVERSITY, NIJMEGEN
PROF ROY FRIEDEN | UNIVERSITY OF ARIZONA
DR STEFANO GOTTARDI | EINDHOVEN
DR MICHIEL VAN HAARLEM | ASTRON
PROF AMINA HELMI | UNIVERSITY OF GRONINGEN
PROF CHARLES H. LINEWEAVER | AUSTRALIAN NATIONAL UNIVERSITY
PROF THANU PADMANABHAN | INTER-UNIVERSITY CENTRE, PUNE
DR SIMONA SAMARDJISKA | RADBOUD UNIVERSITY, NIJMEGEN
DR JOHANNES SCHEMMEL | UNIVERSITY OF HEIDELBERG
DR SALOME SCHOLTENS | UNIVERSITY OF GRONINGEN — UMCG
DR ALESSANDRA SILVESTRI | LEIDEN UNIVERSITY
PROF PETER SLOOT | UNIVERSITY OF AMSTERDAM
PROF LIEVEN VANDERSYPEN | TECHNICAL UNIVERSITY DELFT
DR JOERI VAN DER VELDE | UNIVERSITY OF GRONINGEN — UMCG
PROF ERIK VERLINDE | UNIVERSITY OF AMSTERDAM
DR MANUS VISSER | UNIVERSITY OF AMSTERDAM
PROF ANDERS YNNERMAN | UNIVERSITY OF NORRKÖPING, SWEDEN

FOREWORD
by Robbert Dijkgraaf

The birth of the Age of Information

was a modest affair. On a summer day in 1949, the American engineer and mathematician **Claude Shannon** pencilled a list on his notepad. It was the first attempt to organize the world according to its information content. At the bottom he wrote one bit, the elementary building block, the answer to a simple yes-no question. At the top of the page he placed one hundred trillion bits, Shannon's estimate of the total information in the largest library in the world, the Library of Congress in Washington, D.C. Going down in powers of ten, he subsequently placed everyday items, like one hour of film, one hour of television, the Encyclopaedia Britannica and a gramophone record. Then, somewhere in the middle, just below one hundred thousand bits, he inserted "genetic constitution of man."

That was four years before Watson and Crick discovered the molecular structure of **DNA** and its role as the universal carrier of information in life. By the way, Shannon vastly underestimated the amount of data in a single human cell. Our unique genetic information measures around 3 billion base pairs. Life employs incredibly efficient molecular storage devices. One gram of **DNA** can contain all the current digital data in the world.

This book is in many ways a vastly extended version of Shannon's one-page blueprint. It carries us all the way to the total information content of the Universe. And it bears testimony of how widespread the use of data has become in all aspects of life. Information is the connective tissue of the modern sciences. From its more obvious role in computer science and data-intensive disciplines like climate science and astrophysics, it has now also infiltrated biology, regarding life as the ultimate information processing machine. Even in my own discipline, the search of the fundamental laws of nature, quantum information

Shannon's notebook page of 1949 on which he pencilled the powers of 10, from 10^0 to 10^{13}.

Credit Mary E. Shannon, contained in the book by James Gleick, The Information: *A History, a Theory, a Flood* (Pantheon Books, 2011)

seems at this moment the best candidate for the elementary building block out of which everything can be construed — matter, radiation, gravity, and even space and time itself. One of the great scientific paradoxes, the **black hole**, is from an information perspective both the most simple and complex object in nature. In gravity it is literally nothing more than a hole in space, but in quantum theory it is the most efficient storage device that nature has available. All the data on the internet could easily fit into a black hole much smaller than the known elementary particles.

Will the information paradigm last? Although science aims to be a universal language, the metaphors and images that scientists use are very much a sign of the times. It is no coincidence that the pioneers of the scientific revolution in the seventeenth century described the universe as a gigantic clockwork, just as these timekeeping machines became part of everyday experience. Similarly, at the height of the industrial revolution the human body was imagined as a huge factory with its pipes, valves and turbines, including the brain as management at the top floor.

Undoubtedly, future generations will look back at this time, so much enthralled by **Big Data** and **quantum computers**, as beholden to the information metaphor. But that is exactly the value of this book. With its crisp descriptions and evocative illustrations, it brings the reader into the here and now, at the very frontier of scientific research, including the excitement and promise of all the outstanding questions and future discoveries. Onwards and upwards to the next metaphor!

Robbert Dijkgraaf is Director and Leon Levy Professor at the Institute for Advanced Study in Princeton, NJ, and Distinguished University Professor at the University of Amsterdam.

Shannon
DNA
Black hole
Big Data
Quantum computers

ACKNOWLEDGEMENTS

I would like to express my sincere thanks to Springer Nature Executive Editor Angela Lahee for believing in the concept of this book, and guiding it through production. I also would like to thank Dr Margot Brouwer for assisting me in the demanding and intensive editorial work, Maddalena Munari for the extensive and detailed work on image and copyright as well as proofreading and Ann Scholten-Sampson and Carmen Hoek for support and proofreading. I am indebted to Josje Kobès – Studio Frontaal – for her enormous dedication to making my lay-out concepts come alive.

I have been inspired by and am grateful to my collaborators both for contributing to these pages and for attending the Information Universe conferences at Infoversum/DOTliveplanetarium in Groningen (NL).

Most of my "in vitro" work described in this book has been done with the wonderful Astro-WISE, Target, OmegaCEN, Dotliveplanetarium and Facts and Fakes teams which I was privileged to lead over the years. My endeavours in Target and the Information Universe have been enabled by the fantastic support and trust by the late Koos Duppen, vice president of the board of the University of Groningen 2006-2011.

I thank my wife Marinka and my children for everything.

CONTENTS

CONTENTS

INTRODUCTION

BY EDWIN A. VALENTIJN

E. A. Valentijn, *Powers of Two*,
https://doi.org/10.1007/978-3-030-58345-3_1

JOYRIDING THE UNIVERSE

Although today astronomers are often front runners in using and processing extremely large datasets, the Incas long ago built one of the first 'hard disks' linking the in vivo Information Universe of the Sun and Earth, to the in vitro technological Information Universe.

"Where are we?", "where do we come from?", or better "what is It?". Fifty years ago I decided to become an astronomer, inspired by the Westerbork Synthesis Radio Telescope. The radio data would bring me into deep space and help answer big questions.

My father (a poet who eventually became an accountant) inspired me to contemplate *the* big questions, and my mother as an early feministic editor and writer showed me the benefits of maximal engagement. Growing up in the roaring 60s, I learned to call everything into question and since then I have been joyriding the Universe, in many different fields of astronomy, and at various observatories and international organizations. The more I changed fields and environments, the more I appreciated the commonalities, which must have some relation to **It**. One outstanding similarity is the role of information in nearly everything. It appears as if everything is built of information. Today, both in physics and in our information driven society a most profound question re-emerges: **"It from bit?"**

Eventually, I became a professor in astro-informatics at the Kapteyn Institute, University of Groningen, pioneering and building from scratch Big Data information systems, which for me functioned as a laboratory of a Universe of information. We started with handling the huge data rate of the **OmegaCAM** astronomical wide field imager at the European Southern Observatory (ESO) at Paranal (Chile). Around the year 2000 its 260 megapixel camera superseded every sensor device, and together with the VISTA telescope it delivers an order of magnitude (10x) more data than the sum of all the other existing telescopes of the ESO optical observatories.

I had been there before, when I published with Andris Lauberts the first-ever large set of 32,000 digital images of galaxies in the late 80s (**DVD**). In those days we also had to start from scratch. But this time it was different: it was clear that it would only be a matter of time before the whole world would be swamped by Big Data. With a team of five senior astronomical data engineers we started to design new abstractions from basic principles, scalable and as close as possible to *how nature works*. We evaluated many different approaches, and after two years, in 2003, we celebrated our first "virtual light" in our **WISE** computer systems in which we entangled (linked) all data items.

Today, these **entanglement** abstractions are used for handling the observations from several astronomical observatories such as OmegaCAM, the MUSE imaging spectrometer at ESO's Very Large Telescope and the **LOFAR** radio telescope. While OmegaCAM's telescope was heavily delayed, I got an enormous grant (**Target**) to apply our astronomical techniques to other disciplines, ranging from medical cohorts and PET scans to handwritten text recognition with Artificial Intelligence. Eventually, the **WISE** abstractions were adopted for the **Euclid** astronomical satellite, planned to be launched in 2022. Euclid will image and determine the distances of at least 1 billion galaxies with Hubble Space telescope-type spatial resolution.

In all this work, we built an *Information Universe* in our computer systems (**in vitro** — sometimes called *in silico*). Above all, the ultimate challenge is to understand how this relates to the information in the real physical world around us (**in vivo**). Computer information systems work with **entanglement**, which links and copies information; we brought this concept to the extreme in **WISE** by linking our results to the source of data: the observed phenomena, which is nature. The power of such **links** reflects the mysterious aspect of **entanglement** in our Universe. This entanglement can be compared to the entanglement in quantum physics, and even to current theories by Erik Verlinde of dark matter and 3D entangled **dark energy** in our Universe.

HOW THE POWERS OF 2 WORK:

$2 \times 2 = 4$ and $2 \times 2 \times 2 = 8$. We can write this as $2^2 = 4$ and $2^3 = 8$, in words: "two to the power of two" and "three".

In this book we go all the way up through multiplying with two. E.g. multiplying by two for 33 times leads to $2^{33} = {\sim}8 \times 10^9$.

As every 2 corresponds to the two **states** a **bit** can take (1 or 0), this number corresponds to the ~8 billion unique states a 33-bit string (as at the top of this page) can take.

When making metadata, one 33-bit string could provide a unique address for ~8 billion other bits.

For the total information content of a digital system we are used to think in bytes, which by definition contain 8 bits. Thus the 8×10^9 states of the example can correspond to $8 \times 10^9 / 8 = 1$ gigabyte, the typical content of a DVD.

In this book we go to larger and larger strings of bits, eventually to $2^{399} = 10^{120}$, indeed, a one with 120 zeros.

THE INFORMATION UNIVERSE
Enjoy the ride on the Powers of Two

Here in this book we will take you on a ride through the Information Universe by presenting typical cases illustrating various Powers of Two, starting from the beginning and ending at the universe as a whole.

It is a trip through the world of information, which might guide us in understanding basic questions such as: What is the role of information in our Universe? Does the information have a deeper meaning? Is our universe one big information processing machine? If so, can we understand some of the most basic properties of our Universe? How does the existence of dark matter and dark energy fit in? Are there fundamental differences between information systems such as we build in our Big Data computer world (the Information Universe **in vitro**) and the information contained in nature (the Information Universe **in vivo**)?

I do not pretend to be able to give answers to all these questions, but the endeavor by humans in computer information systems (**in vitro**) can be taken as a laboratory for studying the role of information in nature (**in vivo**).

For this book, I collected some long-lasting notions in the Powers of Two storyline, each pointing to the role of information at various levels, both **in vivo** and **in vitro**. The story starts @ the beginning of our Universe, the Big Bang and the origin of life, then traveling to the world of human beings, who created technologies and complex information systems, to finally go into deep space, back to that beginning. All this can be expressed in Powers of Two. We, as people, are situated in the middle and you, reader, can actually start reading this book from whichever part you prefer. **The keywords of the various story lines are highlighted like this** and also listed in separate boxes next to the text. In the e-version of this book one can click on these links to follow each story line.

This book is certainly not meant as an encyclopedic overview — there is so much more. I have collected items which struck me during my joyride of the universe.

What is the role of information in human **consciousness**? A very intriguing question, as the answer is still largely unknown. I have limited this discussion to the notion of the importance of agreements made by humans (e.g. **ASCII**) enabling the connection between the computer world and the consciousness. I also address **Facts and Fakes**, a topic which actually applies to the scientific endeavor per se. We accept things as true because we believe them to be true given some rules we have made ourselves (the **System of Science** — Hegel), often driven by some kind of consensus made by the **consciousness**. We are not ready for assessing the info consciousness yet. When, 25 years ago, I brought up this question to the cosmologist Andrei Linde he answered: "I cannot handle this topic as long as I cannot understand what it is to feel pain". We have not made much progress since then.

Interestingly enough, computer systems like **WISE** are used to map dark matter in the Universe (**lensing**) and also to test theories of the nature of gravity and dark energy, one of them being Verlinde's information theory of **gravity as an emergent force**. Our tests, done in collaboration with Verlinde, form an ultimate Information Universe **in vivo — in vitro** reflection. While the jury is still out regarding the validity of Verlinde's theory, a theory of this kind, with information as a key building block of the universe, might lead us to understand **It**. Many of the questions are still open, and a lot of scientists are working on them. My journey through the Powers of Two opened many doors, at least for me; I hope you will discover them yourselves while reading these pages.

Welcome to "Powers of Two — The Information Universe" and enjoy the ride!

POWERS OF TWO

The Powers of Two work with three scales in parallel:
- **n: the number of bits in a single string**
  from 0 to 399
  e.g. n = 33 bits
- **the corresponding number of *states* or unique values = 2^n**
  e.g. $2^{33} = 8 \times 10^9$ states
- the volume of data in bytes for which unique addresses can be identified with that string = $2^n /8$
  e.g. 1 gigabyte.

The Powers of Two simply follow 2^n, also called **base-2**, while we humans like to work with fingers, called decimal or **base-10** — 10^n. In computing, the number of bytes is simply base-2 divided by eight.
See the Appendix, page 164, for a table.

Here, in this box, we highlight **keywords** of the various storylines throughout this book

In vivo
In vitro
States
Bit
Base
ASCII
Consciousness
System of Science
Lensing
Astro-WISE
Facts and Fakes
Emergent Gravity

While **Apoptosis**, the ability of cells to kill themselves, is known as a key process of **multicellular** species that only survive when a precise set of signals are received from the environment, University of Alberta researchers found the first evidence that bacteria can cause single-celled algae to kill themselves — a discovery that holds promise for developing newer, more precise antibiotics and producing biofuels.

A bloom of algae called *Emiliania huxleyi* is visible from space off the coast of Newfoundland.

Credit Jacques Descloitres, MODIS Rapid Response Team, NASA/GSFC.

WHAT IS INFORMATION?

What is Information? The simplest answer is: we don't know. This is like other very fundamental elements and fields in our Universe, such as an electron or a photon. It is very difficult to understand these on a basic level. Actually, our understanding emerges from detailed phenomenological and theoretical descriptions of the properties of things. This is also the case for Information, and the meaning of the ride through the Powers of Two is to assemble these insights. Perhaps the most fundamental property, which we will see re-appearing on many of these pages is: *"Information only manifests itself, or even exists in relation to its environment by being copied"* — for example: within the ear information is copied in many different ways a dozen times. Hence Information and *Communication* Technology (ICT), hence the reading of the Intihuatana stone by the Incas in Machu Picchu, hence the reading of the genes of cells by transcription to RNA — which culminates in what is called apoptosis, being one of the most revealing discoveries of the last century. Apoptosis is the built-in selfdestruction mechanism of a cell which is activated when it does *not* receive a copy of a very specific set of the signals from its environment. This is why one cell develops into a liver and another one into a toe nail, while they both contain identical DNA in which all the specific signal sets are stored.

This information aspect being a key question for understanding life, but also cancer, was noted by Piet Borst, Research Director of the Netherlands Cancer Institute, long before the discovery of apoptosis.

Information is technically best expressed as the sum of all the possible states of a system. A bit has two states (e.g. on — off); dice have six states etc. In longer strings of bits such as a 7-bit ASCII character, these bits work together to create a $2^7 = 126$-state system.

Machu Picchu
RNA
Being copied
The ear
Apoptosis
DNA
Bit
ASCII
Multicellular life

*The number of bits in a system corresponds to the number of **states** the system can have as the power of two.*

1-bit corresponds to $2^1 = 2$ states — on and off

CHAPTER 2
THE
BEGINNING
BY EDWIN A. VALENTIJN

© The Author(s), under exclusive license to Springer Nature Switzerland AG 2021

E. A. Valentijn, *Powers of Two*,

https://doi.org/10.1007/978-3-030-58345-3_2

VERY FIRST POWER

The very first power of two: 2^0, corresponds to the value 1. This identifies the single, eternal, indistinguishable state: the primordial sea from which our Universe emerged — sometimes called the Spacetime foam. I call this **Ti**, the reverse of It (from John Wheeler's: **It from bit**). This is one of the miraculous notions in the story of the Powers of Two.

Ti — Spacetime foam

Before the creation of our Universe, information did not exist. No bits, no particles, no life, no agreements. Often people question me:

"What was there before the Big Bang?" or, "Where is our Universe expanding in?".

Cosmologists refer to it as a primordial sea of eternal spacetime foam in which small quantum fluctuations result in new universes, some of them with such odd properties that they collapse and die instantly after their creation.

Our Universe has the property that it contains information. Its amount increases forever during its expansion. Actually, this increasing information corresponds to the number of **states** and can be expressed in increasing Powers of Two, from 2^1, 2^2, 2^3 to up to 2^{256} and beyond that in special theories. However, if we go backwards to the origin of our Universe and again follow the Powers of Two, the Universe started from 2^0 which is equal to one! That is truly remarkable: our Universe did not emerge from nothing (0) but from one (1) — an indiscriminate, undistinguishable unity or state, a totally flat ocean of everything and nothing. Note though, Padmanabhan argues in this book that **CosmIn** carries **4π** information from **Ti** to our Universe. There is nothing on Earth like this — perhaps this picture of a doldrum, which I took from La Palma (Canary Islands), resembles this state of nature. Ocean sailors, stuck in doldrums for weeks, might experience this eternal unity, which eventually gets disrupted by a distant butterfly who triggers a new storm, like the **Big Bang** in our Universe, or alternatively the **Big Melt** in Padmanabhan's views.
Credit NASA/WMAP Science Team

$$Ti = 2^0 = 1$$

Doldrums on the oceans occur mostly around the equator and are calm areas of no wind and consequently no waves, which can last for weeks and trap sail-powered boats in a flat ocean.

From Ti, the primordial spacetime foam, countless universes arise with widely different characteristics: the Multiverse. The Anthropic Principle is a philosophical consideration which states that we, people, will find ourselves in a universe that is suitable for intelligent life to emerge.
Credit 6222336 Algol | Dreamstime.com

MULTIVERSE
Anthropic principle

So-called multiverse theories[1][2] predict that out of the spacetime foam (Ti) many different universes can be created. They could have all kinds of properties, more or less at random. Most of these universes won't last long. Some might last a while but might have a weird structure of space, which looks like spaghetti. In the multiverse theories, our Universe would be just the one which has the specific properties, such as a flat three-dimensional space, that could host the particles, photons and fields all carrying information required for the evolution into humans. This is called the Anthropic principle: universes are created randomly, but we live in the one which could host us — if the universe would not have these specific properties we would not be there. This is why "it is what it is". This is a sobering thought — it neutralizes the "why" question, but it should not stop us from addressing the "how" questions.

Our Universe has the property that space expands with ever increasing chaos (entropy — a measure of information in disorder), while at the same time structures (order) are formed with ever increasing complexity and structured information — an Information Universe per se.

Ti
Anthropic principle
Multiverse
Entropy
Complexity

1 bit
2^1 = 2 states
0 or 1, on or off
Example: 1

At the Big Bang the first bit is created. From the indistinguishable unity of the primordial stuff Ti, "the zeros were separated from the ones": the first bit corresponds to two possible states. This bit is the first step on our journey to capture the ever increasing complexity and amount of bits of our expanding Universe.

"God is an eJ — not a DJ or a VJ" (with the e from email) show at the Noorderzon performing arts festival 2011 Groningen, premiering MicroSoft Kinekt connected to Microsoft World Wide Telescope. Here, in my show, God navigates with his hands through the Universe, explaining the universe of information as observed by astronomers in the Big Data era.

Big Bang

From the single state unity of the spacetime foam, at the moment of the **Big Bang**, the unity separated into two **states** and then beyond, with emerging particles, information and bits. When Einstein said "God does not throw dice" then for the Big Bang the notion is: *"When God created the Universe she separated the zeros from the ones".*

At the Big Bang the Universe goes ON, but from that point on it could go OFF. In other words, the Universe starts with two states: the bit is created. If we take the source of our universe, spacetime foam, as **Ti**, then the bit is created from unity: **bit from Ti**.

The whole process of creation from unity to hydrogen and helium atoms, photons, and forces took less than three minutes. However, the formation of structures has been ongoing for 13.8 billion years with an ever increasing amount of information (**entropy**), chaos and complexity. The combination of an ever-expanding universe of increasing disorder, and at the same time increasing structure and **complexity** (order) is one of the most profound miracles of our Universe. Every time a complex structure is built and order is created, so much heat is generated that the total amount of information and chaos in the Universe increases. This is expressed by the second law of thermodynamics.

Pioneered by J.A. Wheeler, scientists work on physics in which the bit is a fundamental building block of our Universe: hence the view **"It from bit"**, It being here our deepest understanding of the observed universe. Combining these concepts we arrive at "It from bit from Ti".
*Picture on the right **Credit** NASA/WMAP Science Team*

bit from Ti
It from bit
⟶ It from bit from Ti

Information only exists in relation to its environment by **being copied**.

For the traffic sign: copied from the light to the brains of the viewer.

FOUR STATES

The number of bits in a system corresponds to the number of states the system can have:

Two traffic lights = two bits

**Two lights on or off: Four states
(red red, red green, green red, green green)**

WHAT IS A BIT?

A **bit** is anchored in the state of a certain substance and takes one of the two **states** of that thing, for instance a red traffic light that can be *on* or *off*. The two states of the traffic light *on* and *off* only mean something when they relate to the outside world. For humans this often means that we assign a meaning to the two different states: on and off, stop or go, left or right, true or false, etc. We call such a dual entity a bit. The bit, being the smallest unit of information, only gets a meaning in relation to the outside world — which is actually creating **entanglement**.

Likewise, we make agreements among ourselves when we build computers: we build switches in transistors and agree which state of the switch corresponds to a zero or a one. The same goes for storing data in the memory of a computer or any digital storage device: every time a human has provided the meaning of the states, for example whether the magnetic field is up or down on the molecules of a magnetic tape.

For digital signals, the voltage of the internet signal travelling through a cable varies above and below a certain value agreed again by humans: higher voltages correspond to one and lower voltages to zero. The same holds for the way the light signal is modulated in optical fibres.

Basically, humans have introduced definitions to create and interpret bits — and as a next step we build systems obeying these definitions which subsequently support and reincarnate themselves with increasing **complexity**. For example, eventually computer chips are designed and produced by using computers. This Information Universe is built by humans, like a huge laboratory: *the Information Universe* **in vitro** (sometimes called in silico). As some minimum level of consciousness or adaptive response of a being is required, the in vitro Information Universe is as old as these beings — are these the **multicellular organisms**? But nature does the same on its own, without any need for humans. The Big Bang is the beginning of the **in vivo** Information Universe 13.8 billion years ago.

Credit Ingram Publishing

2 bits

$2^2 = 4$ states

MULTICELLULAR SPECIES

Any form of life is based on *exchanging* information between cells. This started 2 billion years ago when multicellular species emerged. By exchanging information, cells collaborate and act as a unified whole: life — starting with a very simple form of intelligence.

Amoeba — unicellular

Credit Flinn Scientific

Multicellular Life
In vivo

Billions of years ago unicellular systems evolved into multicellular ones: current estimates range from 2 – 2.5 billion years ago for algae species to 570 – 750 million years ago for respectively animal fossils and frond-like creatures. The momentous transition is believed to have happened many times for different species and likely over and over[3].

In any event, from that moment on cells started to exchange information, which is equal to copying information. This exchange of information can be seen as the start of **life** and its evolution, beginning with dedicated cells to monitor light, sound — the senses — to detect food, to survive and to corporate between cells. Actually, life started with the exchange of information, and information in turn only exists when it is **being exchanged / copied**. Complex life started originally with **multicellular life** forms exchanging information internally!

We would tend to think that the start of life marks the start of the in vivo Information Universe, but this is not true: the particles and photons of the early Universe already carried and exchanged information for billions of years before multicellular life started. You can watch them on the screen of an analogue TV set with an antenna on your roof: 13.8-billion-year-old photons of the **Cosmic Microwave Background (CMB)** copying into your TV set.

The neural networks in the **human brain** form an ultimate instantiation of cells exchanging information, in other words "communicating cells". This is called cell signalling — see pages 40-45 by Lineweaver.

Multicellular green algae Tetrabaena socialis
Credit Hisayoshi Nozaki and Yoko Arakaki

**Multicellular life
Being copied
Human brain
CMB
Life**

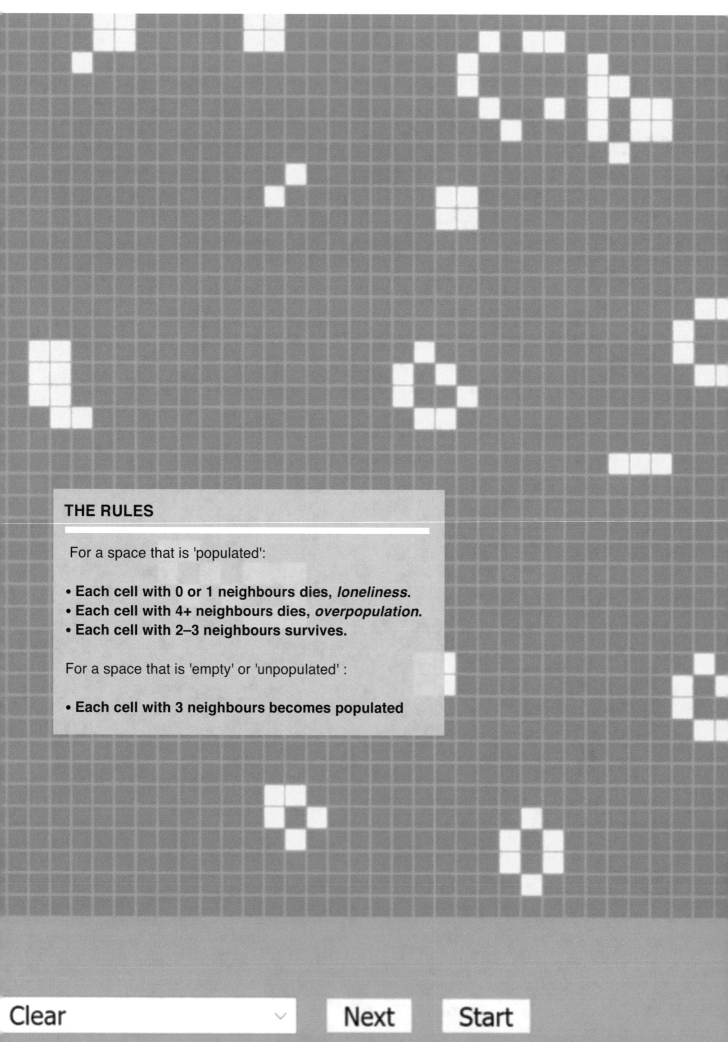

THE RULES

For a space that is 'populated':

- **Each cell with 0 or 1 neighbours dies, *loneliness*.**
- **Each cell with 4+ neighbours dies, *overpopulation*.**
- **Each cell with 2–3 neighbours survives.**

For a space that is 'empty' or 'unpopulated' :

- **Each cell with 3 neighbours becomes populated**

Clear   ∨   Next   Start

THE GAME OF LIFE *In vitro*

Cambridge mathematician John Conway's Game of Life ran already on the first simple Personal Computers in the early 70s. A quite simple set of computer instructions created visuals on the screen with apparently living, growing, moving and dying objects and organized structures. Indeed, with only a few lines of code already working on the first MS-DOS machines. It became a cult success and many versions followed. But the key suggestion was that, apparently, basic properties of life and everything in the Universe can be created from information in a 'cellular automaton' — the in vivo – in vitro cross-over. It is seen as a demonstration of "nature as an information processing machine". The first time I learned about it is from Gerard 't Hooft demonstrating it as a case for an information layer in physics represented by cogwheels.

Credit Edwin Martin

John H. Conway, a renowned mathematician who created one of the first computer games, dies of coronavirus complications — April 2020.
Credit Courtesy Diana Conway

Cross-over
In vitro
In vivo

Most of the human activities in the Information Universe take place in the domains of power 1–64 (2^1 – 2^{64}). Human DNA and brains are in the middle of the range with respectively power 32 and power 36. Modern computer technology reaches the power 64 domain and beyond, involving much more bits than contained in the human brain — though it takes an enormous amount of energy compared to the brain.

In the 2020 decennium astronomers will push information technology to new limits, with their so-called Stage IV cosmological observatories such as the Euclid satellite, the SKA radio telescope and the Rubin telescope, involving exa-(10^{18}) and zetta-(10^{21}) bytes of data. The Stage IV suite of cosmology projects will collaborate to unveil the most pressing mysteries of our Universe.

## CHAPTER 3

PEOPLE'S INFORMATION UNIVERSE

BY EDWIN A. VALENTIJN

and contributions by:

Charles H. Lineweaver
K. Joeri van der Velde
Salome Scholtens
Katrin Amunts
Johannes Schemmel
Alessandra Silvestri
Amina Helmi
Anders Ynnerman
Margot Brouwer
Heino Falcke
Lieven Vandersypen
Michiel van Haarlem
Simona Samardjiska
Peter Sloot

7 bits
$2^7 = 128$ states
16 bytes

ASCII LIST

There is currently no physical theory explaining how the digital world connects to human consciousness. In the world of Information Technology (IT) all information exchange is based on agreements between people. For example the **ASCII** list connects the human mind's alphabet to a 7-bit string.

Every touch on the keyboard generates an **ASCII** character string — every letter on the screen in turn is based on this.

ASCII
in vitro

Information exists when it is **being copied** — a rather important notion — think of bits, **DNA** and lottery. This concept is highlighted by **Shannon's** classical paper on Information theory which describes a mathematical theory of communication, communication being equal to the copying of information. The second leading notion is that the copying process only works when agreements have been made. Actually, everything we do with computers is based on agreements among people — often in the form of standards.

Our **in vitro** Information Universe is fully based on agreements between people. Agreements are actually mostly tables in which we connect items. These connections are rather spooky and are actually created in the **human brain** (or by the human who instructed computers). These are called **links**, associations or references — different names for the most profound aspect of the Information Universe.

For complicated information systems, achieving these agreements is often a major obstacle, and involves an incredible amount of project management and "sociology" to align not only standards but also behaviours, protocols and ways of doing things. Eventually, this is all built in **data models**.

Globally, the process of settling **standards** for computing is nearly fully controlled by the big companies which, in an open world, win the competition for adaptation of standards, e.g. word.doc, adobe.pdf. Sometimes it is even controlled by enemy buy-outs of the competition. An exception is the **ASCII** standard.

Dec	Oct	Hex	Binair	Code	Betekenis
97	141	61	1100001	a	Letter a
98	142	62	1100010	b	Letter b
99	143	63	1100011	c	Letter c
100	144	64	1100100	d	Letter d
101	145	65	1100101	e	Letter e
102	146	66	1100110	f	Letter f
103	147	67	1100111	g	Letter g
104	150	68	1101000	h	Letter h
105	151	69	1101001	i	Letter i
106	152	6A	1101010	j	Letter j
107	153	6B	1101011	k	Letter k
108	154	6C	1101100	l	Letter l
109	155	6D	1101101	m	Letter m
110	156	6E	1101110	n	Letter n
111	157	6F	1101111	o	Letter o
112	160	70	1110000	p	Letter p
113	161	71	1110001	q	Letter q
114	162	72	1110010	r	Letter r
115	163	73	1110011	s	Letter s
116	164	74	1110100	t	Letter t
117	165	75	1110101	u	Letter u
118	166	76	1110110	v	Letter v
119	167	77	1110111	w	Letter w
120	170	78	1111000	x	Letter x
121	171	79	1111001	y	Letter y
122	172	7A	1111010	z	Letter z

ASCII character code

LYNDON B. JOHNSON
XXXVI *President of the United States: 1963-1969*
127 - Memorandum Approving the Adoption by the Federal Government of a Standard Code for Information Interchange.
March 11, 1968

top The **ASCII** character code is a table associating the alphabet to a 7-bit string — The alphabet being the language of people, the bits being the units in computers. In the end, the table builds the direct connection between our computers and our **consciousness**. We are using it all the time without knowing it. In fact, the table represents a list of **links**.

above While in the open world standards are set by the financially strongest companies or strong groups of dedicated volunteers and scientists, a notable exception is the decree by the President of the United Stated in 1968: *"All computers and related equipment configurations brought into the Federal Government inventory on and after July 1,1969, must have the capability to use the Standard Code for Information Interchange".*

Being copied
DNA
Shannon
In vitro
Human brain
Links — addresses
Data Model
Standards
ASCII
Consciousness

Machu Picchu

COLLECTION OF 8 BITS

A collection of 8 bits is called a **byte**. There is nothing fundamental in a byte — it is a practical definition, just to have a natural (Power of Two) unit which is close to the alphabet (7-bit) and the number of human fingers (decimal numbers).

On my return from an observing trip to ESO's telescopes at la Silla Chile in the early 80s, I made a trip to **Machu Picchu** in Peru. Such a magical place, on a very steep mountain surrounded by green valleys. The magic culminated at the top: when spotting the Intihuatana stone it became immediately clear to me that here the Incas would watch the shadows made by the Sun and perhaps some stars, to spiritually and practically interpret the heavens. I was thrilled when I took my pictures, as somehow the stone ultimately connects in a very basic way the **in vivo** Information Universe made by the Sun and the stars and the **in vitro** Information Universe made by humans — the ultimate astronomer's experience.

Indeed, archeologists state the stone was used as an astronomical clock to determine particular moments during the year: the equinox (start of spring autumn) when the Sun is right above the stone not making any shadow, and the solstices (start of winter/summer). This information was then used to plan agriculture. The rays of the Sun (**in vivo**) beam onto the stone's surfaces which work as on/off switches made by humans (**in vitro**), altogether determining dozens of states of the system.

That the stone contains 8 bits corresponding to 1 byte is my personal guess; it could well be the Incas used more surfaces, then the stone contains more bytes. In any case, one could rightfully say that the stone represents a first **hard disk** with at least 1 byte — 8 bits — of memory made by the surfaces.

The Intihuatana stone, a giant rock carved by the Incas of ancient Machu Picchu in Peru, can be considered as a first 8-bit hard disk. Why so? The surfaces point to the cardinal directions marking the start of the four seasons. As the Sun's rays lit the different surfaces of this huge rock throughout the year, it triggered the Incas activities: sowing, harvesting, celebrating and praying.

This ancient stone dissolves both the boundaries between heaven and Earth, and those between the digital and natural Information Universe. In fact, the stone represents an ultimate picture of the cross-over between the in vivo and the in vitro

Information Universe. In vitro means the man-made technology to handle information and in vivo means the information built in nature, in this case the orbit and the light rays of the Sun.

The Inca priests were astronomers and considered the Sun as a supreme natural god; the priests' reading of the Sun affected societal life and agriculture.

16 bits 2 bytes
2^{16} = 65,536 states
8 kilobytes

First Computers

1970

When computers emerged in the 1970s, astronomers first adopted them to steer their telescopes. Back then, a maximal effort to understand the mathematics of the problem was needed to squeeze the solution into the small computer memory. Nowadays, this is all very different.

The Intel 8086 chip was launched in 1978. It was the first 16-bit microprocessor and started the revolution of personal computers at home.

Credit Chipdb.org

A 16-bit number can have 65,636 different values (**states**). That is a sufficiently large number to be able to instruct computers to do interesting things, or better, to compute interesting functions with enough precision. It all started off in the 70s with the PDP-1 and IBM-360 machines that could work with respectively 18- and 16-**bit** (2-**byte**) numbers (words). These mainframes were not accessible to the public, which dramatically changed in the late 80s with the introduction of personal computers fitted with the Intel-8086 16-bit processor.

As usual, astronomers were early adopters of new IT and the Westerbork radio telescope, opened in 1970, synthesized radio maps of the Universe through a nearly fully computerized handling of the signals.

As a student I determined serious misalignments of the telescopes by taking pictures of the night sky with a photographic camera mounted on a telescope. The memory bank of the computers that steered the telescopes across the sky was too small to adjust for these misalignments without knowing the cause. Then, after understanding the misalignments, we fixed the problem by reformulating it in a few terms in the steering computer.

I went through a similar experience when commissioning in 1984 the INT optical telescope on La Palma; the telescope was wobbling seriously and was steered by an old Perkin Elmer 16-bit machine. This was typical for our work in those days, using maximum understanding to squeeze the problem into the limited computer. Today, particularly in the domain of **Artificial Intelligence**, scientists and programmers often do the reverse.

above The official inauguration of the Westerbork Synthesis Radio Telescope by Her Majesty Queen Juliana of The Netherlands — here in the helicopter — took place on June 24, 1970. This event took place a few days before the legendary Holland Pop festival, Kralingen Rotterdam: the first 3-day (love, peace, music) festival in the Netherlands, following the US Woodstock festival. I made analogue recordings. Bootleg records were produced in these days following an ideology to provide more open and affordable access to music, which was eventually realized by todays streaming services on the internet.

Credit The Rise of Radio Astronomy in the Netherlands: The People and the Politics—by Astrid Elbers

right IBM Personal Computer, commonly known as the IBM PC. IBM model 5150 was introduced in 1981. It was created by a team of engineers and designers under the direction of Don Estridge in Boca Raton, Florida.

Credit Rama & Musée Bolo

States
Bit
Byte
Artificial Intelligence

16 bits   2 bytes
$2^{16} \sim 6 \times 10^4$ states
~7.5 kilobytes

Which can in turn produce
$2^{60,000} = 10^{17,962}$ states

BATS

Can one virus-carrying bat paralyze the whole world? Certainly yes, as the bat can be viewed as the person-zero of the pandemic. But also no, as there are trillions of possible viruses and only a very small fraction makes it: this fraction is also strongly determined by, and thus effectively mirrors, the environment of the bat.

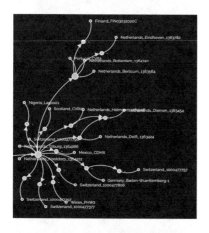

MORE TO COME

It has been claimed for a long time that viruses top the natural disaster list, that there will be many more to come with their own identity — and that some of their properties will mirror our society. Unlike another high on the list: Earth being hit by Near Earth Objects, which my group is studying for ESA.

COVID-19

The Coronavirus is an **RNA** virus consisting of single-stranded RNA of approximately 29,834 base pairs, which corresponds to about 7.5 kilobytes of data on your hard disk. How can it be that just a single 7,458-**byte** information unit is hitting the whole world? These 7.5 kilobytes correspond to less than a 1,000th of a second of information displayed on your TV and around a second of music on your radio. How can such a small amount of information possibly have such a huge impact? The trick is that the string of around 30,000 base pairs, which contains two bits of information per base pair, act as a single unity. *So it does not act according the number of possible configurations (or states) of a single 16-bit string, instead it represents a string of length 60,000 bits.* This can be viewed as an enormously long address or barcode. In fact, this barcode can identify a truly unmeasurable amount of *unique* **addresses** (sometimes called **states** in this book): $10^{17,962}$ — much more than needed to address each single particle in the Universe. So, there is an incredible amount of *possible/potential* viruses of this RNA makeup, but most of them will never become a virus, not able to make an envelope, to replicate or to find an environment which can host them. So by means of just a random lottery of spontaneous mutations, it is only a matter of time for one of these possible viruses to become the one that is able to survive and fits to a hostile environment. Indeed, the surviving virus will consequently mirror the society which hosts it — per se. The virus itself is not clever. The truly clever trick is the ability of nature to glue the base pairs into long strings.

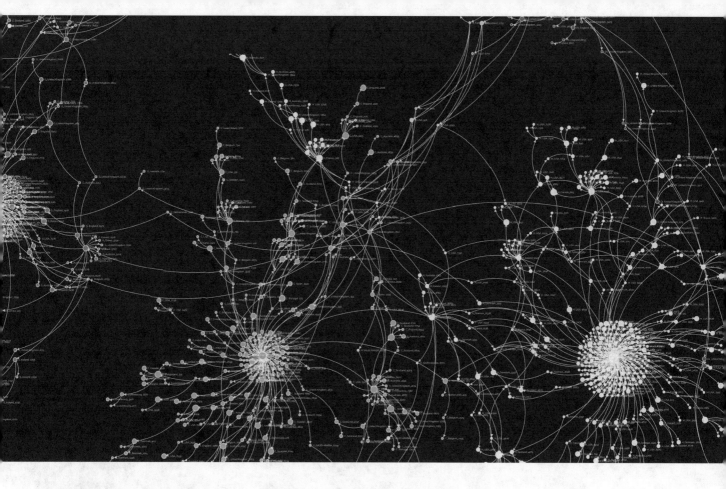

INCONVENIENT TRUTH

In chaos theory (James Glieck[4]) it is recognised that the flapping of the wings of a butterfly could cause a hurricane 1,000 miles away, provided it just happens at the right moment and place. This can be true, but its only a part of the story, as after all the anti-cyclonic winds of the hurricane are caused by something enormously bigger: the rotation of the Earth around its axis. Likewise, one could recognise the COVID-19 outbreak as caused by a single animal, possibly a bat, carrying the COVID-19 virus, which paralysed the whole world. This can be true as well, but after all the global 2020 outbreak is facilitated by something much bigger: an over-population of the world with nearly unconstrained economic objectives and its worldwide commuting — a problem which I was taught at school more than 50 years ago — apparently an inconvenient truth.

STATISTICAL STUDIES

Statistical studies estimate the occurrence of about 320,000 viruses in mammals[5]. But also a "single" virus, like COVID-19, mutates continuously due to errors in the copying (replication) process. The evolution or mutation tree of COVID-19 virus as gathered and displayed by Graphen shows hundreds of mutations during only a couple of months' evolution.

Credit Graphen

RNA
Links — addresses
States
Life
Chaos
Byte

STAR PEACE

MACHU PICCHU AT THE CANARIAS

29 June 1985 — The inauguration of the Observatorio Roque de los Muchachos La Palma, a collection of many telescopes. The largest collection of royals ever at an observatory and also the first in Spain after the Franco era. The event was headed by King Juan Carlos, striving for unity and collaboration in Europe and beyond. In his impressive speech he highlighted "the harmony of galaxies" as a source of inspiration for collaboration and friendship among the nations of Europe. This was in the days that Ronald Reagon was launching his "Star Wars program". The guests of honour included their majesties King Juan Carlos and Queen Sophia of Spain and their son and two daughters, Queen Margrethe and Prince Hendrik of Denmark, Queen Beatrix and Prince Claus of the Netherlands, King Carl Gustaf and Queen Silvia of Sweden, the Duke and Duchess of Gloucester. President and Mrs Hillary of the republic of Ireland and Dr and Mrs von Weizsacker of the Federal Republic of Germany. The event was attended by hundreds of ambassadors, ministers and journalists and was broadcasted live for four hours on the Spanish national television.

Credit RGO- David Calvert

"Edwin Valentijn saved the life of the Dutch Queen Beatrix by catching her just before falling over the crater's edge at the inauguration on La Palma" according to the headlines in Dutch newspapers. Fake news stories are at all times alike and can only be dispelled by tracing links of information to their source, links or associations being a fundamental property of the Information Universe. I was most surprised that this headline was copied in dozens of newspapers and became more and more exotic. In the end "the royals were panicking and almost tumbled into the Caldera". No one bothered to check the source of information. The root of fake news is not in the internet (it did not exist yet) — it is in the minds of people. This event, among others, inspired me to set up a Facts and Fakes tracing research programme recently. Background picture: the inauguration of the Kapteyn telescope, where I am surrounded by the Spanish royal family. In the foreground Queen Beatrix and my PhD advisor Prof Harry van der Laan.
Credit NFP

Facts and Fakes
Links — addresses

PRE-INTERNET *FACTS AND FAKES*

Security was very high — the ETA terrorist group was a serious threat and King Juan Carlos was surrounded by guards who would not hesitate to put their bodies in the line of fire. Terrorist attacks in the 70s and 80s in Western Europe were much more eminent and killed many more people than today. Most believe it became worse in the last decade — but this difference is one of awareness and is mostly due to the rise of internet and communication media in the 90s. Todays' terrorism is dominated by fundamental Islamic-inspired jihadists, while in the 70s and 80s there were unconnected ideologies. In a Dutch PhD thesis the candidate has to present a number of statements — one of mine in 1979 reads: *Press, radio, TV and other media contribute enormously to the effect of terrorist attacks.*

I wish I was wrong.

Credit Datagraver.com 2020, *Data* START GTD
Credit RGO, David Calvert

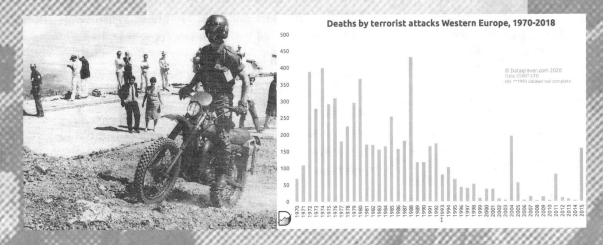

Deaths by terrorist attacks Western Europe, 1970-2018

© Datagraver.com 2020
Data: START GTD
NB. **1993 dataset not complete

24 bits   3 bytes

$2^{24} = 1.7 \times 10^7$ states

2 megabytes

FIRST MEGABYTE STORAGE

The first megabyte storage devices formed a stepping stone towards an Information Technology era which today is 4 billion times more efficient today.

A 2.5-petabyte IBM storage unit — TS1150 (2019)

Credit International Business Machines Corporation (IBM)

Hard disk

Only 60 years ago, a 5-megabyte **hard disk** weighed over 5 tons, and had to be loaded onto an airplane by a truck. Today, we carry 10,000 times more information in a trouser pocket. This demonstrates the amazing advancement of information technology over the past decades which, also for storage media, seem to follow **Moore's law** originally formulated for the advancement of processors in computers: *the performance of computers doubles every two years*.

In 2019, a single hard disk carries up to 31 terabytes of data: ~1,000 times more than a typical mobile phone, and ~6 million times more than the IBM hard disk from 1956.

In today's data centers, high-performance and large storage capacity is obtained by integrating many parts together in a unit and by bluntly putting many units in long rows. In 2019, IBM offered an integrated storage solution consisting of eight units of the physical size of the 1956 hard disk, but instead containing many tape drives which together provide 20 **petabytes** (20×10^{15} bytes) of storage. This is 4 billion times more than the 1956 hard disk of the same company.

Amazingly, this result is very close to the expectation of Moore's law if applied to 64 years of development. With 1,000 of these units one could store full DNA scans of the whole world population. I do not think this is a good idea, but the truly astronomical numbers demonstrate we are really building an Information Universe in our machines.

A 5-megabyte IBM hard disk loaded onto a plane weighed over 5 tons (1956).

Credit International Business Machines Corporation (IBM)

Hard disk
Moore's law
Petabyte

THE TELEPHONE

IN VIVO – IN VITRO

As a precursor of the internet, the telephone offered many of the same advantages and dangers, and was heavily discussed at its introduction. Whether telephone or the internet, it all revolves around communication or copying of information. The telephone was one of the major discoveries of the 20th century.

Credit The U.S. National Archives

LIFE IS INFORMATION

Life is information. Human **DNA** contains about 3×10^9 base pairs which roughly corresponds to the number of states that can be described with a 32-bit number and stored on a 500-megabyte drive. How is this information transferred?

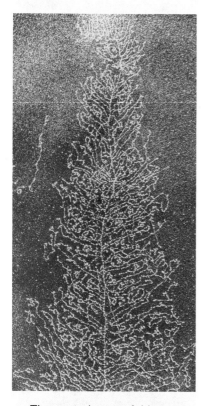

The central stem of this "tree" is mRNA. The "branches" coming off the stem are proteins getting longer as ribosomes (See page 42) work their way down the mRNA.
Credit ANP I Oscar Miller/ Science PhotoLibrary

32 bits   4 bytes
2^{32} = 4×10^9 states
0.5 gigabytes

DNA
by Charles H. Lineweaver

How much information is in Human **DNA**? Inside the cells of our bodies are coiled strands of DNA (Deoxyribo-Nucleic Acid). When DNA in a chromosome is uncoiled and magnified, its structure as a double helix or a twisted ladder can be seen. The rungs of the ladder are nucleobases: adenine (A), cytosine (C), guanine (G) and thymine (T), paired in four possible ways: A-T, T-A, C-G, G-C.

If we imagine all the DNA inside one of your 50 trillion cells as one long twisting ladder, the ladder would have 3 billion rungs made of 3×10^9 base pairs. The **information density** of DNA, in terms of the separation between rungs, is 3.4×10^{-10} meters, making the total length of the DNA molecule ~1 meter.

If a miniature molecule-sized you climbed down the ladder and looked at one side of the ladder, you would see a series that would be something like this: ACGTGTAATTCGTGTCCAT-TAACTACTGGCCTTATCGTGTGTCAATCGACTTGAGATAG-GCATATCTCAA...

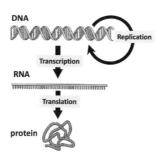

"Information only exists when it is being copied" is a recurring theme in this book. DNA is exemplary. Information is trans-ferred by means of complex biochemical processing with a key role for **RNA**.
Credit Gwen Childs

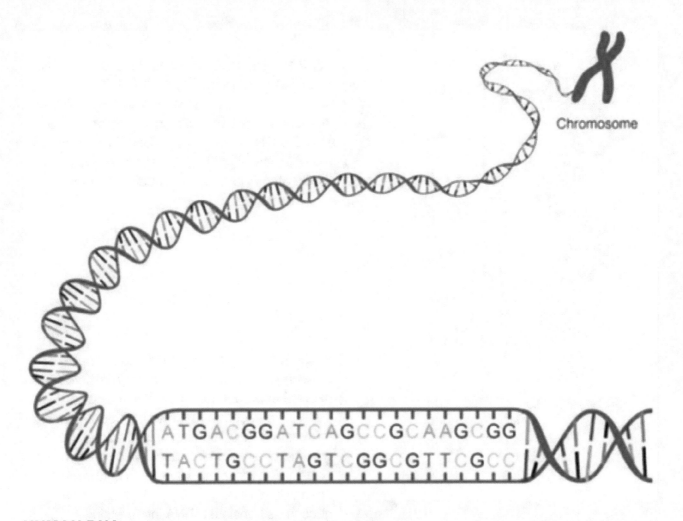

ATGACGGATCAGCCGCAAGCGG
TACTGCCTAGTCGGCGTTCGCC

Chromosome

HUMAN DNA

How can we quantify the information in DNA? A stretch of DNA that is only eight base pairs long might look like this: CCGAAACG. There are four possibilities in each position: either A, C, G or T. Thus, DNA is a **base-4** information system, and not **base-2** like bits in the Powers of Two. We can represent the bases with the numbers zero, one, two and three. The largest number is then 33,333,333 and the lowest number is 00,000,000. However, unlike numbers, one cannot say that A is greater than C, or T is greater than G. So, talking about larger or smaller numbers doesn't make sense. They are more like **addresses** or **links**. Including zero, an eight-digit **base-4** coding system can represent $4^8 = 65,536$

different configurations or states. Human DNA is not a sequence of eight bases. It is a sequence of $N = 3 \times 10^9$ bases. Thus, the DNA in one of your cells can specify

$$4^N = 4^{(3 \times 10^9)} = 10^{(1.8 \times 10^9)}$$

different states. By recognizing that $2^2 = 4$ we can write:

$$4^N = (2^2)^N = 2^{2N} = 2^{(6 \times 10^9)}$$

So, 6×10^9 bits (750 megabytes) of information could distinguish all possible base-4 DNA sequences of length 3×10^9. Only about 1% of our DNA gets translated into protein. If we only want to quantify the information

that gets translated to protein, then in the equations above we need to replace N with N/100.
Credit DNA Strands via Wikimedia Commons

DNA
Information density
RNA
Base
Links — addresses

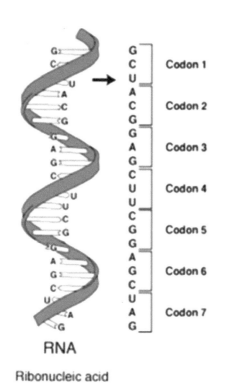

RNA

Ribonucleic acid

G
C
U — Codon 1
A
C — Codon 2
G
G
A — Codon 3
G
C
U — Codon 4
U
C
G — Codon 5
G
A
G — Codon 6
C
U
A — Codon 7
G

STEP 1 Transcription: Consider a sequence of nine DNA base pairs: TGG AAA GAT. The first step in the transfer is when it is transcribed into Messenger **RNA** (mRNA) and becomes UGG AAA GAU. Notice that when DNA is transcribed into RNA, thymine (T) is replaced by uracil (U). *Credit* Sten André via Wikimedia Commons

STEP 2 In the second step the ribosome takes in Messenger **RNA** (mRNA). Various Transfer-RNAs (tRNAs) approach the ribosome. If a specific tRNA's triplet codon matches the codon of the mRNA, that tRNA's amino acid is added to the growing chain of amino acids which becomes a protein. The most important parts of ribosomes are made of RNA. This video illustrates how ribosomes translate mRNA into protein: www.youtube.com/watch?v=Jml8CFBWcDs *Credit* Boumphreyfr via Wikimedia Commons

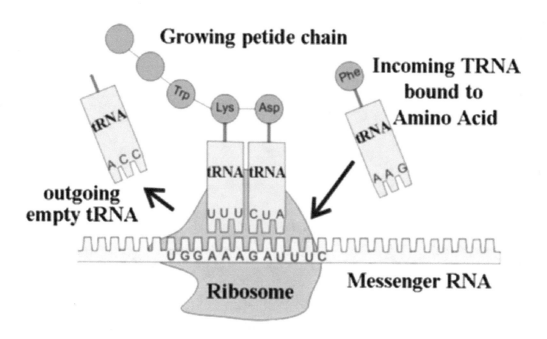

Growing petide chain

Incoming TRNA bound to Amino Acid

outgoing empty tRNA

Ribosome

Messenger RNA

FROM 1D DNA TO 3D PROTEINS

BY CHARLES H. LINEWEAVER

How is the one-dimensional digital information stored in **DNA** transferred into the three-dimensional information of the proteins of your body?

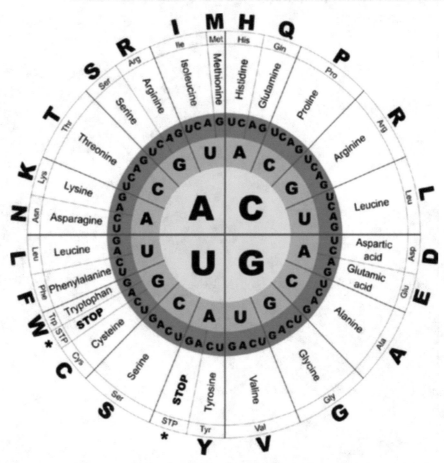

STEP 3 The wheel above shows the universal code that ribosomes use to translate mRNA codons into amino acids to form proteins. Consider the sequence UGG. This figure tells us which amino acid this sequence codes for. The first base is U, so begin with the large U in the center of the circle. The second base is G, so we move to the smaller G at about eight o'clock. The third bases are the smallest letters. Moving to the smallest G we see that UGG codes for the amino acid tryptophan (Trp). In the diagram of STEP 2, the amino acid Trp has just been released by its tRNA and added to the growing chain of amino acids (pink). **_Credit_** Reproduced from Open Clip Art

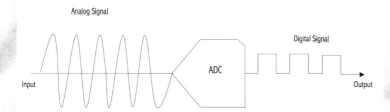

Analog Signal

Digital Signal

Input

ADC

Output

INFORMATION ONLY EXISTS IN RELATION TO ITS ENVIRONMENT BY BEING COPIED

In the language of information processing, the architectural information of the environment is 3D and analogue. The biological information in DNA is 1D and digital. Thus, Darwinian evolution can be understood as an environmental-to-biological, 3D-to-1D, analogue-to-digital converter.

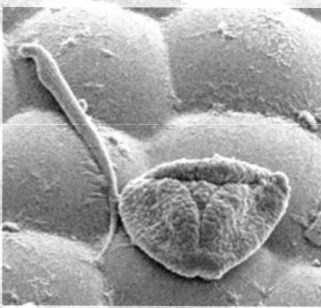

The background image is an electron micro-scope photo of the compound eye of an ant. The approximately hexagonal components are omatidia (clusters of photoreceptor cells). The triangular hat-shaped object is a pollen grain.
***Credit* Charles Lineweaver**

***Credit* Carl Spitzweg**

WHERE DOES BIOLOGICAL INFORMATION COME FROM?

BY CHARLES H. LINEWEAVER

Life forms are packed with information. Consider the three-dimensional information in your fingers. The information to build those fingers (and every other part of you) is stored in the one-dimensional information of the 3 billion base pairs of DNA in each of your cells. Where did all this biological information come from? How did it get inside the DNA? The simple answer is that it came from the environments in which our ancestors evolved. Environmental information is the origin of biological information.

Before there was life — before there was any biological information — the environments on Earth were complicated and full of architectural information: in the bright Sun it was hot and dry; it was cool and moist underneath the rocks. There were hot places, cold places, dry places, wet places, sunny places and shady places. Conditions varied between morning and evening, and cycled between night and day, high tide and low tide, summer and winter. We are not sure how the first life forms got started, but they needed an information-based ability to survive and reproduce within these complicated conditions. If their cell walls were too permeable, they died. If their cell walls were too impermeable,

they died. If their skin was too thin, they died. If they couldn't find water for their young, their young died and then they died. If they couldn't outrun a predator they died. If they couldn't hide in the bushes, they died.

The death of many and the survival of some, is how information about the environment gradually diffuses into the DNA of the survivors. Life forms did not "figure out" how to survive a gauntlet of rivals, predators, parasites and viruses. During this gauntlet, most creatures die. The survivors are, therefore, no longer a random sample of those who started. The surviving sample has been skewed towards being more able to survive. This shift in the surviving population is the result of the transfer of environmental architectural information into biological information. Natural selection, the central engine of Darwinian evolution, has introduced a correlation between the environment and the surviving biological equipment that was used to survive in that environment. In the Information Universe, Darwinian evolution is how information gets transferred from the architectural information of the environment to the biological information in the DNA of the survivors. This information-transferring gauntlet never stops.

DIGITAL REVOLUTION

It's amazing how fast and far the digital revolution has come since 1989. Around 30 years ago Philips lab approached me since they had made "a big discovery": it was possible to store a lot of digital images on a CD! They were chasing me for digital images, since I had 32,000 galaxy images. NASA at that time had less than 1,000.

The first optical storage media appeared in 1986 and were developed by Philips and Sony. The ESO-LV image collection was the first to be stored on optical media and was migrated a dozen times to next generation optical media, until these were phased out by storage in the cloud.

Credit Chronozoom

33 bits
2^{33} = 8.6×10^9 states
1 gigabyte

DVD

Adriaan Blaauw, director general of the European Southern Observatory (ESO), polled all the astronomical institutes of the ESO member states in the mid-70s for their interest in mining ESO's new Quick Blue photographic plates for galaxies. As there was not much extragalactic optical astronomy in Europe in those days, the only respondent was Erik Holmberg from Sweden. He assigned Andris Lauberts to the task who did an incredible job of it: he *visually* inspected with a microscope all 407 of the 30x30 centimeter plates and compiled a catalogue of 16,000 ESO-Uppsala galactic images[5].

The human-made entries in the catalogue — nowadays we would call these annotations — enabled us to automate the digital scanning of three times the area of these galaxies with a 20 micron single-spot aperture PDS micro densitometer. Scanning whole plates was out of reach in those days. The scanning was done in the '80s on both the Quick Blue and Red ESO survey plates, leading to a room full of magnetic tapes and the first large set ever of 32,000 digital images by 1988 which I published together with Lauberts — ESO-LV[6].

Eventually, I was called by Philips Lab when they had just discovered that they could use the CD music technology to record images on CD-I optical disks — I for interactive. But there were no digital images that could fill the disks. NASA had fewer than 1,000 — we had 32,000. When I told this story around the year 2000, hardly anyone could take in that this was only ten years before, illustrating how fast the digital imaging revolution took place. The CD-I was a commercial failure of at least 1 billion dollars and stopped in 1996 when overtaken by **DVD** multimedia devices.

Scanning machines like ESO's plate scanner PDS mark the transition from analogue to digital. With this single spot scanner photographic plates of the skies were digitized. The data were stored on magnetic tapes, which in the case of the ESO-Uppsala Survey filled whole rooms with tape racks in the eighties.
Credit ESO.

In those days astronomers were already aware of the enormous amount of data contained in the universe, and the limits of our resources on Earth — sometimes we used to rest our case by quoting *"The sky is the best archive"*. But then the digital revolution took of, and positivism on handling very large datasets emerged:

I remember the late Jim Gray (research director Microsoft) stating at the momentous **Virtual Observatory** conference[7] in Garching bei München in 2002 *"all our present day large datasets would fit in the spare space of future generations of storage media"*. He turned out to be quite right. At the same conference, I presented how the **Astro-WISE** information system deals with large datasets as living distributed archives. On the other hand, positivism about collecting enormous amounts of data up to peta- and exabytes prevailed in astronomy, which led Alexander Szalay — who worked closely with Jim Gray — to present a talk at the first Information Universe conference 2015 entitled: *"How to collect less data"*.
Credit DVD-multimedia, Philips

DVD
Virtual Observatory
Astro-WISE

35 bits
$2^{35} = 3.4 \times 10^{10}$
48.000 individuals genotyped
at 1,000,000 markers

DEEP LEARNING

Deep learning helps us estimate which DNA mutations are likely to cause disease, and which are harmless. Still, it is like searching for a needle in a haystack.

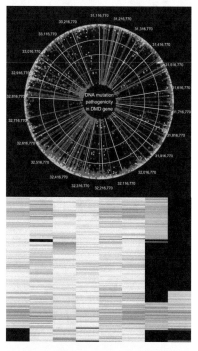

There are many ways to associate genes to diseases, for instance by literature mining or molecular networks. We see that both methods add unique pieces to the genetic puzzle.
Credit Joeri van der Velde et al.

Biobanking for human health

BY K. JOERI VAN DER VELDE & SALOME SCHOLTENS

How can the tiny DNA molecule give rise to Big Data, and how could that benefit our health? What is the role of biobanks and algorithms in the future of medicine? DNA sequencing and other cutting-edge molecular technologies have brought us to the point where doctors can detect inborn diseases before they even are manifest. But how? We have discovered nearly 180,000 locations in our genome that influence how our bodies are formed and how healthy we are. This includes general traits like hair colour, height, word reading ability, and freckling — but most traits are of medical relevance and include susceptibility to common diseases such as hypertension, arthritis, celiac disease, cancer subtypes, diabetes, cardiovascular disease, inflammatory bowel disease, obesity, allergies, psoriasis and asthma. Genetic mutations may increase or decrease risk for these diseases, prompting personalized check-ups and lifestyle advice. When treating, pharmaco-enomic passports inform patients which drugs should be taken in higher or lesser doses to be the most effective. In addition, clinical genetics has identified around 4,000 genes causing severe rare disease when sufficiently perturbed. Although individually rare, together rare diseases are common: around one in seventeen people suffer from a rare disease, many of whom children. Genome diagnostics unveils the patient-specific gene mutations that cause rare disease to offer the best help for patients and families. Essential for the success of molecular medicine are biobanks. They enable research into common and rare diseases by collecting biological samples and data from both healthy individuals and patients. Biology is transformed into Big Data by DNA sequencing and other modern technologies. Within these digital collections, researchers discover the intricate relationships between DNA and health using advanced algorithms such as deep learning. These relationships drive the translational research of developing new tools for clinical practice. Together, we are shaping the medicine of the future by going from *in vivo* to *in vitro* and back again.

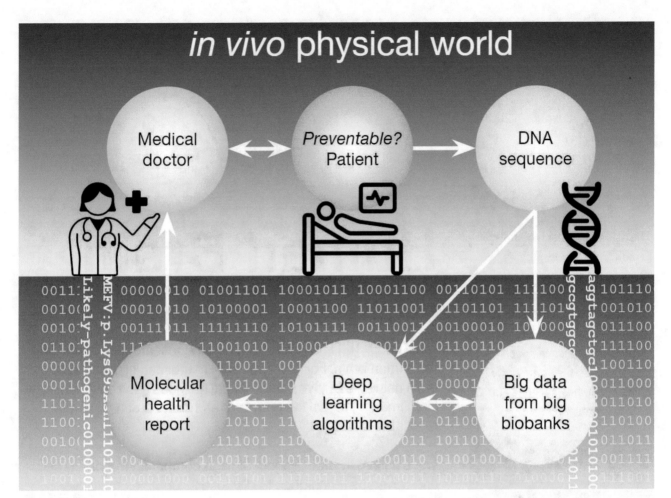

in vivo physical world

Medical doctor ⟷ *Preventable?* Patient → DNA sequence

Molecular health report ← Deep learning algorithms ⟷ Big data from big biobanks

lifelines

**167,000 participants
3 generations, 30 years**

top Underneath modern molecular medicine is a vast 'digital world' of data and algorithms. The DNA can be sequenced from a patient or better yet: a healthy individual with a preventable disease. Information infrastructure such as MOLGENIS then places the DNA in the Big Data context of everything we have learned from biobanks and research. Together, they enable algorithms to provide a molecular diagnosis, prognosis, preventive measures, personalized treatment, family planning advice, and hopefully one day, repair of the underlying genetic defect.
Credit Joeri van der Velde with embedded credits aan Noun project

above Lifelines is a large, multi-generational cohort study that includes 10% of the northern population of the Netherlands. Hospital data, detailed questionnaires, blood samples and **DNA** profiles have been collected over the last ten years, making Lifelines the largest genetic Big Data biobank within the Netherlands. By measuring the DNA, gene expression, metabolites, microbiome, lifestyle, and disease, we can find out how one affects the other — which diseases are caused by nature, and which by nurture?
Credit LifeLines B.V.

DNA
Big Data
Lifelines

36 bits
$2^{36} = 7 \times 10^{10}$ states
9 gigabytes

BRAIN COMPLEXITY

The human brain has nearly 10^{11} neurons. If we assume that each neuron can have just two states, on or off, this would result in a gigantic number of putative neuronal states. Neurons form networks at different scales in space and time. This **complexity** is one of the challenges of today's research.

This page is dedicated to Karlheinz Meier, who died in October 2018 — here at one of his last presentations at the Information Universe Conference 2018.

Human Brain
BY KATRIN AMUNTS & JOHANNES SCHEMMEL

Different methods allow researchers to look into the human brain. For example, with neuroimaging one captures brain networks **in vivo**, while optical methods such as Polarized Light Imaging reveal the network's architecture at high spatial resolution **in vitro**.

The EU Human Brain Project aims at computing the brain, and develops techniques to disclose its complex multilevel organisation using computers, i.e. the rules that underlie these neuronal networks and their function. It is estimated that each neuron has up to 10,000 contacts, i.e. synapses, so in total there are around 10^{15} synapses in the human brain, which brings it to the domain of **petabytes**, **qubits** and quantum computing.

Information processing at the level of cells and their connections is highly complex — neurons have dendrites and an axon, they form networks on different spatial scales, and are modulated by chemical substances. Each neuron is a unique entity, a complicated three-dimensional structure, that exchanges messages with other neurons.

Petabyte
Qubits
Complexity
In vivo
In vitro
Human brain

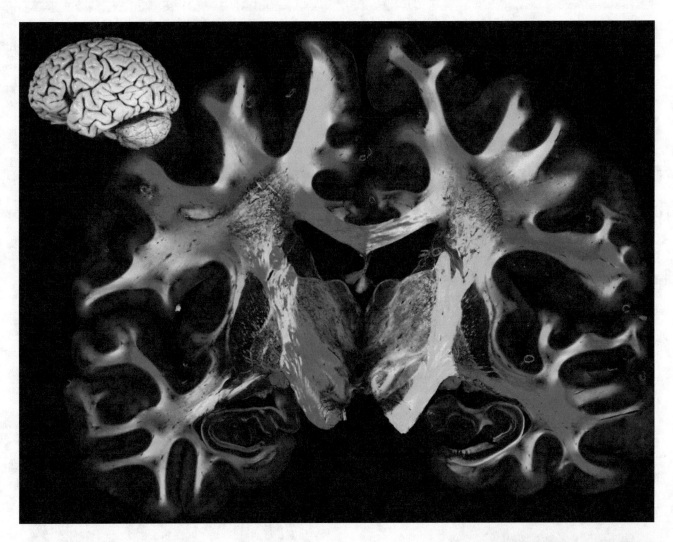

above **Human brain** histological section showing the directions of axons (colour coded) — **in vitro**. The network architecture was visualized using 3D Polarized Light Imaging, a method that is based on special optical properties (the birefringence), of myelin sheets, surrounding axons — Axer et al. 2016[9)].

Myelin contains lipids, which creates this optical effect. It is necessary to isolate axons from each other, and to increase the speed of signal conversion. A 3D reconstruction of a whole brain axonal architecture in a computer requires 2–4 **petabytes** of storage. **Credit** Heidelberg University

right *This cylinder filled with yarn rolls helps you to imagine the many connections that exist between neurons. It has been estimated that a human brain contains about 2–3 million kilometres of "cables", i.e. one would need 200 cylinders like this filled with yarn to model the length of connections of one brain.* **Credit** *Heidelberg University*

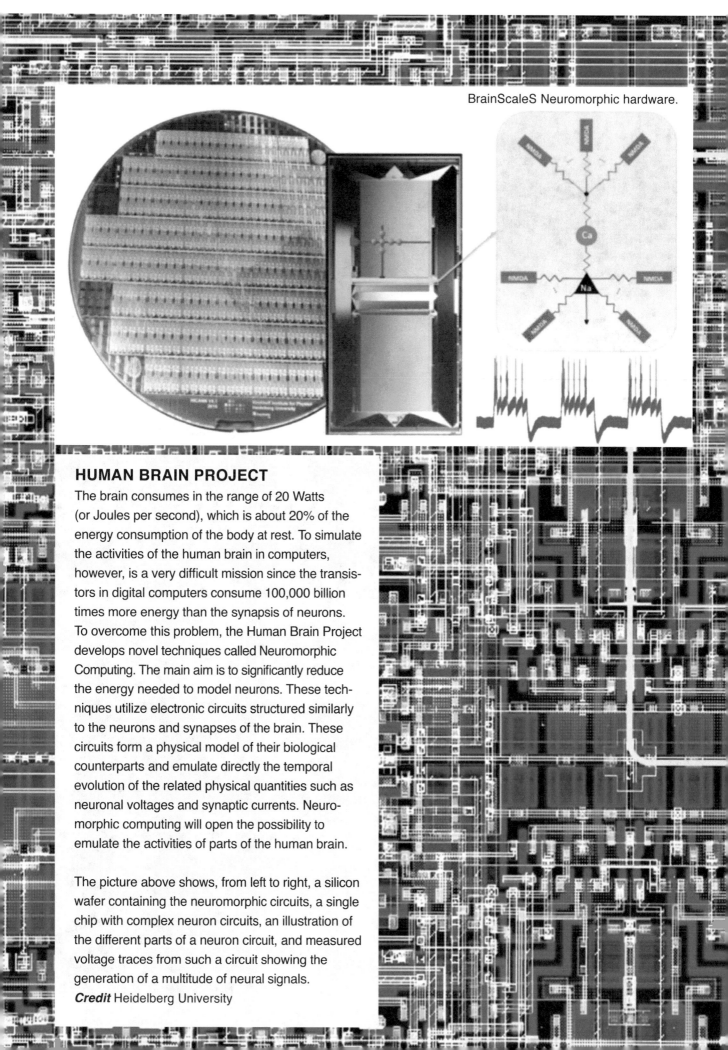

BrainScaleS Neuromorphic hardware.

HUMAN BRAIN PROJECT

The brain consumes in the range of 20 Watts (or Joules per second), which is about 20% of the energy consumption of the body at rest. To simulate the activities of the human brain in computers, however, is a very difficult mission since the transistors in digital computers consume 100,000 billion times more energy than the synapse of neurons. To overcome this problem, the Human Brain Project develops novel techniques called Neuromorphic Computing. The main aim is to significantly reduce the energy needed to model neurons. These techniques utilize electronic circuits structured similarly to the neurons and synapses of the brain. These circuits form a physical model of their biological counterparts and emulate directly the temporal evolution of the related physical quantities such as neuronal voltages and synaptic currents. Neuromorphic computing will open the possibility to emulate the activities of parts of the human brain.

The picture above shows, from left to right, a silicon wafer containing the neuromorphic circuits, a single chip with complex neuron circuits, an illustration of the different parts of a neuron circuit, and measured voltage traces from such a circuit showing the generation of a multitude of neural signals.
Credit Heidelberg University

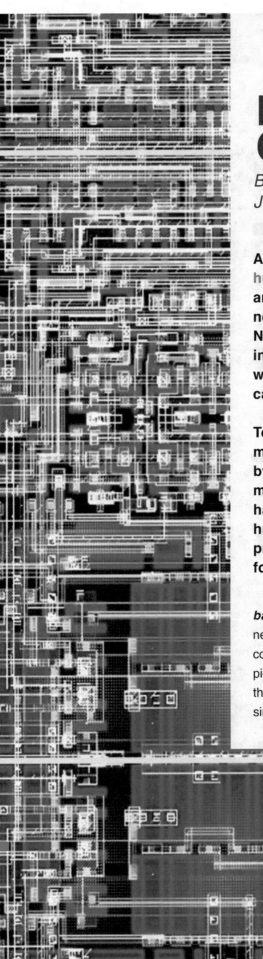

NEUROMORPHIC COMPUTING

BY KATRIN AMUNTS &
JOHANNES SCHEMMEL

A powerful instrument to understand the human brain is to simulate its neuronal network architecture. The Human Brain project emulates networks in neuro-inspired computers — Neuromorphic Systems — to get new insights into brain functions. Moreover, such systems will help to reduce the energy needed for these calculations.

To cover the information content of these messages, imagine a computer with a 9-giga-byte memory changing the whole content of this memory 1,000 times per second — this is what happens in your brain. To better understand the human brain means to understand the manifold processes underlying information processing, for example the energy efficiency of the brain.

background image More than 100 tiny transistors are needed to mimic a single synapse of the brain. Their complicated geometrical arrangement is shown in the picture. A silicon wafer contains close to a 100 million of these circuits — approximately 10% of the synapses in a single cubic millimeter of the brain of a mouse.

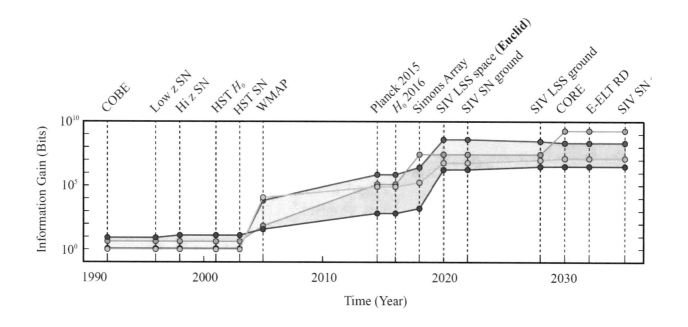

CUMULATIVE INFORMATION GAIN ACHIEVED IN THE PAST AND PREDICTED FOR THE FUTURE, BASED ON THE DATA ACQUIRED BY COSMOLOGICAL MISSIONS

Different lines and colours show the cumulative (i.e. accumulated via successive experiments) information gained on several models (different colours) that go beyond the standard model of cosmology, the ΛCDM model, by either modifying the laws of gravity or introducing **dark energy**.

The units for the information gain are 'bits' of the Kullbach-Leibler divergence which, in information theory, quantifies the knowledge gained over the parameters of a theory with respect to the a priori (i.e. starting) knowledge we had of them.

The horizontal axis also displays future cosmological missions (SIV=Stage IV, the upcoming suite of experiments). From the early 90s to the upcoming SIV programmes this implies an information increase by a factor 2^{34}, nearing the total information content of the **human brain**.

It is evident that we are about to witness a next jump in information that will allow us to probe new physics and discern among alternative cosmological models.

Credit Alessandra Silvestri
Adapted from Raveri et al.(2016), arXiv:1606.06273.

DARK ENERGY— DARK MATTER

BY ALESSANDRA SILVESTRI

Einstein's crucial insight that space and time are themselves players in the cosmic drama provided us with the revolutionary and powerful theory of **General Relativity (GR)**, which has proven extremely successful in describing gravity around compact objects and in the Solar System. It has also provided us with a coherent theoretical framework within which to study the cosmos.

Fast forward a century, and we have now an established model of cosmology, Λ Cold Dark Matter (**ΛCDM**), based on the laws of **GR**. Yet, our understanding of the major constituents of the Universe is limited: only ~5% is made up of component parts that we know! The remainder is in the form of **dark matter** (~25%), responsible for the clustering of matter around us, and **dark energy** (~70%), responsible for the accelerated expansion phase the Universe entered some billion years ago (cosmic acceleration). We still lack theoretical models for the workings of both and yet, *it could simply be that our understanding of gravity on large cosmological scales is incomplete*.

Indeed, all our inferences about the evolution and content of the Universe crucially rely on the assumption of **GR** as the correct theory of gravity. But this represents a huge leap of faith to scale and curvature regimes that are vastly different from those of the Solar System, where **GR** has been confirmed. Could we do without the mysterious dark components if we modified the laws of gravity on cosmological scales? If yes, what is the correct theory of gravity? If not, can we learn more about the nature of **dark matter** and **dark energy**?

The worldwide astronomical community has embarked on an intense observational effort to address this. Up and coming missions, such as ESA-led **Euclid** will offer an impressive information gain on the nature of gravity on cosmological scales. This will allow us to discern among many candidate models, eventually establishing whether we need to extend our theories beyond **GR**. It will also shed light on the nature of **dark matter** and **dark energy** for the latter, the cosmos is our only available laboratory with incredibly high **Information densities**

General Relativity
ΛCDM
Dark matter
Dark energy
Euclid
Human brain
Information density

38 bits
$2^{38} = 3\times10^{11}$ states
~30 gigabytes

GAIA SPACE MISSION

The Gaia space mission has measured the positions and motions of 1.3 billion stars in our Milky Way. Using this data, simulations, and a big dose of common sense, Amina Helmi discovered a galaxy that collided with ours 10 billion years ago and won the Spinoza prize.

Credit Reyer Boxem, Dagblad van het Noorden

When I see a pattern, I need to understand where it came from. I'm convinced that, even when applying advanced data mining tools, you always need to validate the outcome against your own intuition and expectations before drawing a conclusion.

Gaia: a billion stars
BY AMINA HELMI

How do you discover one new galaxy amidst 1.3 billion stars? The information collected by the Gaia space mission is overwhelming in both size and **complexity**. While floating in deep space, the satellite spins around every six hours, continuously observing the stars of our Milky Way. Each star is observed 70 times, so for every object in the data you need to figure out exactly which star you are looking at. From 200 terabytes of raw data, the Gaia consortium extracted the 3D positions and 2D motions of 1.3 billion stars: the largest and most precise 5D space map ever made. For 7 million stars we have the velocities in all directions, giving us 6D information. Fortunately this set of **Metadata** from the survey is 7,000 times smaller than the raw data: "only" ~30 gigabytes, which easily fits on our computers. And, because it is public, even you at home are able to download it (though analysing it is another matter).

Exploring an **Open Science** dataset is a race against time: every scientist wants to be the first to make a great discovery. Through our experience, we knew which combination of observables is interesting, and that made all the difference. Stars from other galaxies which are pulled into our Milky Way should still be moving together in the form of "stellar streams", so we immediately plotted their 3D velocities. Looking at these plots, I remembered a similar pattern in a **simulation** that one of my PhD students made ten years ago. The data turned out to be a perfect match to this simulation of a large galaxy merging with our Milky Way. But just this similarity is not a definitive proof. We used data from another survey, APOGEE, to study the chemical composition of the stars in the structure. What we saw immediately sealed the deal: the composition of these stars was completely different from those born in our Milky Way, proving that they were once part of another galaxy that collided with ours long ago.

Snapshots of a galaxy merger in a **numerical simulation**, where the arrows indicate the positions and motions of the stars. The bottom panels show our Milky Way in white, and the colliding galaxy in red. Thanks to simulations like this, we discovered a large galaxy that was consumed by the Milky Way 10 billion years ago, which we named Gaia-Enceladus. Working with simulations revealed we were on the right track, but also pushed us to look for confirmation in other datasets. Thanks to simulations like these, we acquired the intuition to interpret the real data quickly and correctly. In my view, this shows that human intuition remains a very important part of research, even in this computer dominated era. So far, I have never seen a result from **Artificial Intelligence** that was fully unexpected, and yet made sense afterwards. So unless we can transmit our intuition to computers, it will be hard for Artificial Intelligence alone to make unique and totally surprising discoveries. *Credit* ESA

Complexity
Metadata
Open Science
Numerical simulations
Artificial Intelligence

38 bits
2^{38} = $3{\times}10^{11}$ states
34 gigabytes

FULL BODY SCAN

A high resolution modern CT scanner can acquire data with a spatial resolution of approximately 0.1 millimeter. A full body scan, which only takes a few seconds to complete, then generates up to 20,000 slices of image data, with 1024x1024 12-bit voxels per image in total, corresponding to 2^{38} bits (31.5 gigabytes).

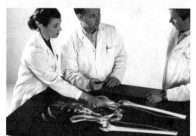

Credit Image courtesy of Sectra AB

Now it is possible to look inside animal and human bodies on touchscreens. Forensic investigations on, for instance, corpses of victims can be done with touch-screen tables. You can look inside, rotate, scroll and zoom animal and human bodies using tens of gigabytes of CT scan data.

CT Scan
BY ANDERS YNNERMAN

Ever since the development of the first **CT** scanners in the 1970s both the speed of the scanner and the resolution have increased dramatically. Using X-ray sources and detectors that rotate around the patient, information is collected from many views in image projection planes as the patient is moved through the gantry. A full body scan can be completed in just a few seconds. Computers then analyse the collected image data in a process called **computer tomography** (**CT**), which determines what 3D object in the scanner caused the 2D X-ray data in the image planes, and results in a volume of data representing the interior of the body.

Each point in the volume is called a voxel and contains information about how much X-ray radiation has been absorbed in that particular point in the body. Different kinds of human tissue absorb different amounts of X-ray radiation and it is possible to separate out different organs and features using this information.

To make images a technique called "volume rendering" is often used. For each pixel on the screen, and 60 times per second, a virtual ray is traced through the volume, which picks up information such as colour and opacity of encountered objects in the data. This tracing is done on graphics processing units GPUs and makes it possible to interactively explore the virtual body.

CT scanning is still under rapid development. We can already look at moving objects such as the beating heart of a living patient inside the scanner or see the interior dynamic function of a knee. Very soon it will be possible to use detectors to count individual photons and generate multi-spectral data for each of the points in the volume, thereby even getting information about the chemistry of the body in a **CT scan**.

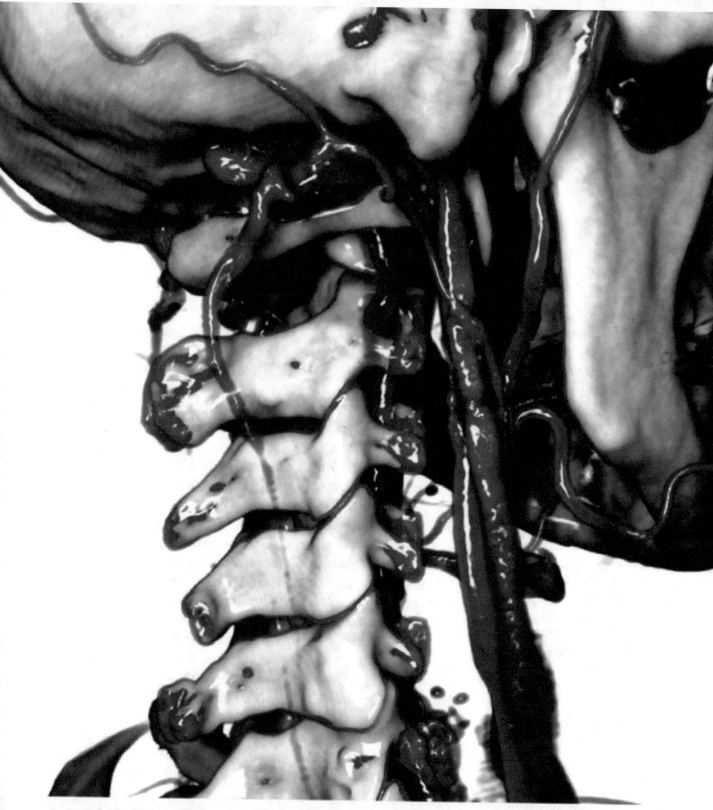

Photon mapping of a **CT** dataset. Virtual photons are allowed to scatter light inside the volume leading to effects such as occlusion shadows and mirrored highlights of the blood vessels.

Credit Data Courtesy: Center for Medical Image Science and Visualization (CMIV)

CT scan

43 bit
$2^{43} = 8.8 \times 10^{12}$ states
1 terabyte

VLT SURVEY TELESCOPE

Dark energy and dark matter: two mysterious constituents of our Universe. How do astronomers get and handle the data from the **VLT Survey Telescope** on a high mountain top in Chile to shed light on these "still too dark" topics? This telescope surveys the sky every hour at night generating terabytes of data.

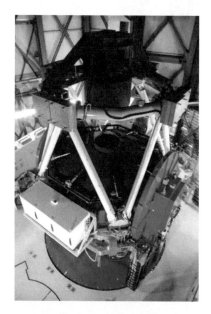

The VLT Survey Telescope (VST) at Paranal, ESO Chile

Terabyte

What is dark matter? The study of the nature of **dark matter** has always been a lead in my work — starting from my PhD thesis on "Radio investigations of clusters of galaxies" in 1978. Already in these early days it was clear that the number of stars and amount of gas in and around cluster galaxies was far too small to bind them together in clusters. The simple existence of these galaxy clusters required dark matter.

Then, around the year 2000, a new window opened with the advancement of the digital CCD imaging devices. By mosaicing 32 CCDs in one camera we built a wide field imaging camera **OmegaCAM** with 260 million pixels. At this time a 1-million-pixel photo camera was state of the art. The wide field camera can be used to capture an enormous portion of the sky in depth, allowing us to stack the images of many individual galaxies to detect the signatures of dark matter by means of the **lensing** technique. A single image is about 0.5 gigabytes — we take pictures every two minutes, so after ten nights we have about a 1,000 gigabytes = 1 terabyte of raw data: the beginning of the **Big Data** era. In 2000 my role was to lead the team that designed and built the data handling and analysis system. We started from scratch and built it with an international consortium: **Astro-WISE**. The information system saw its first virtual light in 2003 in an international data federation. Then disaster struck. The primary mirror of the telescope arrived in 1,000 pieces in Chile, and the project was severely delayed.

Anticipating that it was only a matter of time before Big Data would overwhelm our society, I acquired in 2008 € 32 million for setting up the **Target** project: a grant of € 16 million and the rest contributed by IBM, Oracle, SMEs and public organizations. Target deployed the Astro-WISE information technology to other disciplines and companies — the Information Universe in vitro unfolds in the Big Data era.

OmegaCAM image of the globular cluster Omega Centauri **Credit** G. Sikkema for the OmegaCAM consortium

Broken VST mirror
Credit ESO

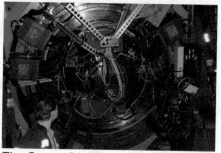

The OmegaCAM camera

Dark matter
OmegaCAM
Lensing
Astro-WISE
VLT Survey Telesco
Big Data
Target

PARANAL OBSERVATORY

In the Atacama desert (Chile) the European Southern Observatory built the 21st century version of Machu Picchu. The four Very Large Telescopes (VLTs) can be seen as very sensitive zoom lenses, while the two smaller survey telescopes Vista (upper left) and the VLT Survey Telescope (next to the VLTs) are wide angle telescopes. At the start of their operations in 2012 they acquired ~10 terabytes of raw data per month, about 25 times more than the huge VLTs together, marking the beginning of the Big Data era.
Credit ESO

Oh leave the Wise our measures to collate
One thing at least is certain, LIGHT has WEIGHT,
One thing is certain, and the rest debate —
Light-rays, when near the Sun, DO NOT GO STRAIGHT.

— Arthur Eddington

GRAVITATIONAL LENSING

BY MARGOT BROUWER

What is gravity? And what does it have to do with war, religion and a trip to Africa? World War I, "the war to end all wars", had just begun when German physicist Albert Einstein opened fire on Isaac Newton's theory of gravity. He proposed that space and time are not static, in fact, their curvature is equivalent to gravity. Of course, nobody believed him; due to the great tensions between Germany and the rest of Europe, only a few astronomers even wanted to consider a theory developed by the enemy. Against all odds, British astronomer Arthur Eddington, a pacifist due to his Quaker religious beliefs, planned an expedition to observe a solar eclipse on the African island of Principe. Not only did this help him to avoid wartime service (which he dreaded), it also allowed him to test Einstein's theory of General Relativity. If Einstein was right and mass curved spacetime, the path of light rays emitted by faraway stars would be bent as they passed the Sun. This could be measured during an eclipse. The position of the stars would appear shifted, an effect called gravitational lensing. Just after the war, Eddington traveled to Africa and measured the right amount of curvature to confirm Einstein's theory over Newton's, making them both instantly world-famous.

Now, more than 100 years later, this same lensing effect is used to perform observations that completely confound Einstein's theory. Just as the light of stars is bent by the spacetime curvature around the Sun, the light of faraway galaxies is bent when it passes a massive galaxy cluster on its way to Earth. The light of these galaxies gets stretched out in a ring-like shape around the center of the cluster, as seen in the background picture of galaxy cluster "Abell NGC 2218". By measuring the distorted shapes of the background galaxies, we can estimate how massive such a galaxy cluster is. The answer: Way too massive! The mass measured through gravitational lensing is about five times higher than what we would expect from the visible light of the cluster: the light of the galaxies and gas. This might mean that all the matter we know is only 20% of the mass in our Universe, while 80% consists of invisible particles which we call dark matter. Alternatively, it might mean that Einstein's gravity, like Newton's, needs to be revised in a way that transcends the old theory, e.g. by using information theory or the holographic principle.

General Relativity

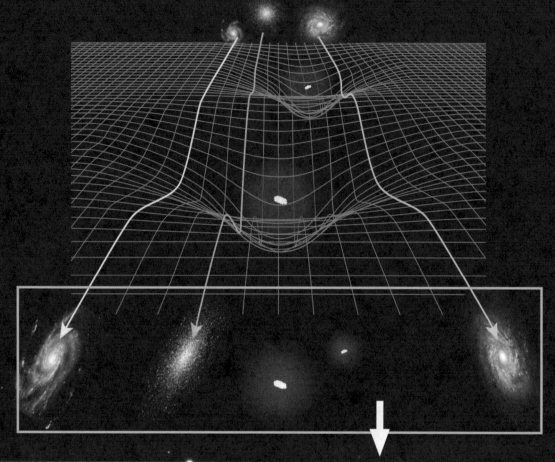

Z_B	ALPHA_J2000	DELTA_J2...	el_A
0.25	128.52013	-1.0297	-0.3011
0.66	128.90613	-1.02934	-0.1818
0.18	129.41364	-1.02916	-0.3941
0.12	129.09045	-1.029	0.3964
0.79	128.69883	-1.02894	-0.0094
0.26	129.34591	-1.02889	0.0358
0.25	128.51876	-1.02868	0.4033
0.23	129.15543	-1.02909	0.0733
0.33	128.68493	-1.02852	-0.0282
0.23	129.33482	-1.02929	-0.1557
0.1	128.9778	-1.02851	-0.5998
0.13	128.63147	-1.02864	-0.4662
0.44	128.96828	-1.02961	0.189
0.32	129.33921	-1.02831	-0.275
0.17	129.04741	-1.02806	0.1917
0.79	128.70015	-1.02788	0.3492
0.39	129.35128	-1.02791	-0.0671
1.24	129.09826	-1.02786	-0.2042
0.7	129.34035	-1.02824	-0.3689
0.89	128.96822	-1.02778	0.4601
0.35	128.51193	-1.02771	0.0819
0.7	128.70502	-1.02769	-0.1674

Ωm

$\sigma 8$

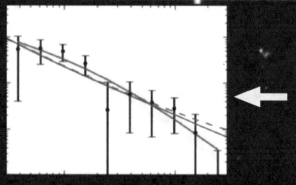

Dark matter
Lensing
Big Data
Emergent Gravity
VLT survey telescope
Metadata
CMB
ΛCDM
Euclid

When I compared the mass distribution of ~33,000 lens galaxies to Erik Verlinde's new theory of Emergent Gravity (EG), I found that the result matched his prediction without the need for dark matter. Whether his theory holds up to other observations remains to be seen. In any case, this result demonstrates an intriguing connection between the amount of normal matter and dark matter in galaxies; and possibly even the Hubble constant, which describes the expansion of our Universe.

Credit Alex Tudorica, KiloDegree Survey . *Credit* APS/Alan Stonebraker; galaxy images from STScI/AURA, NASA, ESA, and the Hubble Heritage Team *Credit* Margot Brouwer et al.2017.

FROM TERABYTES TO TWO NUMBERS

BY MARGOT BROUWER

Understanding the nature of dark matter is one of the holy grails of modern cosmology. Using the gravitational lensing effect, we can measure the true distribution of mass around galaxies, both visible and invisible. But because the force of gravity is so weak, this distortion is incredibly small: only 1% of the *original* shapes of the galaxies, which we don't even know! We need to perform a statistical analysis, using the shapes of thousands of background galaxies, to measure the mass distribution around one single foreground galaxy. And even then, the signal is still too weak. We need to combine the measured mass distributions from thousands of foreground galaxies (or "lenses"), in order to obtain one reliable measurement of the average distribution of dark matter around them. This is why astronomers love Big Data.

To perform these measurements, we use observations from the VLT Survey Telescope. We convert the terabytes of sky image data collected by this telescope into just a few numbers, which describe the distribution of dark matter in our Universe. How is this possible? First, we measure the galaxy locations, shapes and distances from these sky images, creating metadata consisting of gigabytes of galaxy information (already 1,000 times smaller than the original images). Finally, through the statistical lensing analysis, we convert this information into only about ten numbers which describe the mass distribution around galaxies.

We can even convert all our Metadata into just two numbers: Ω_m, the total density of dark matter in our Universe; and σ_8, the amount of clustering of dark matter on an 8-megaparsec scale. Together, these two numbers describe how dark matter in our current Universe behaves. Strangely, scientists find a discrepancy between these numbers as measured locally through gravitational lensing, and measured from the very early Universe through the ancient Cosmic Microwave Background (CMB) radiation. This means our Universe might have evolved in ways we did not expect. If so, our whole view of space, time, and everything within it (given by the ΛCDM paradigm) could be flawed. How will we find out how our Universe truly evolved? What we need is a telescope that can peer deeply into space, to observe the evolution of the cosmic dark matter structure over billions of years. This is the goal of the Euclid satellite.

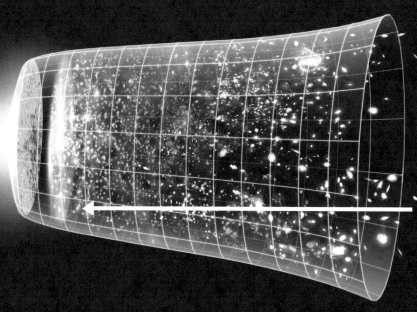

CMB

13.8 billion
years ago

**Local
observations**

Now

EUCLID
Looking 10 billion
years into the past
Credit NASA WMAP
Science Team

Why did the Universe evolve the way it did? Dark matter and dark
energy are the two main players in our cosmic origin story. Right
now, we only have two windows into their nature: the present
Universe as shown by local observations, and the dawn of time
as shown by the Cosmic Microwave Background (CMB) radiation.
Using the ΛCDM standard model of cosmology, we estimate how dark
matter and dark energy evolved over the 13.8 billion years between
these two windows. But discrepancies between local observations
and the CMB are starting to show this model's limits.

Hubble Legacy Field
More than 250 days

The full Moon
Size of one Euclid image
Less than 1.5 hour

To make a 3D map of almost the entire Universe, Euclid must work
very fast. How fast? Almost 5,000 times faster than the Hubble Space
Telescope, but with the same resolution and accuracy. Engineers
and scientists are working around the clock to prepare this
unprecedented instrument for launch at the end of 2022.
Credit NASA, ESA and GG.Illingworth and D. Magee.
Credit NASA/GSFA/Arizona State University

THE EUCLID SATELLITE

BY MARGOT BROUWER

Would you like to travel in time? How about travelling billions of years into the past, to solve the two biggest mysteries in cosmology? This is what Euclid, ESA's new space mission to map the dark universe, is going to do. It will make a 3D map containing the locations of billions of galaxies: a "CT scan" that slices up all of time and space between now and 10 billion years ago. It will grapple with the mysterious dark energy by measuring the evolution of large-scale structure, tracking the expansion of our Universe. Meanwhile, it will study the evolution of the cosmic dark matter distribution through gravitational lensing. For the first time in history, we will know how these dark components evolved over the history of the Universe. This crucial information is likely to shed some light on these two decades-old cosmic enigmas.

Credit ESA

CT scan
Dark energy
Dark matter
Lensing
CMB
ΛCDM
Euclid

$$2^{53} = 9 \times 10^{15}$$

1 petabyte

PETABYTES

How do we deal with petabytes of stored data? The beginning of the Big Data era and the **data-centric** view in information technology.

1,000 books in a billion files.

Credit Image from the diaries of Leo Polak (1901), from the University Library Amsterdam, provided to the Monk system at RUG by Dr Stefan van der Poel.

LOFAR Radio telescope

Credit Netherlands Institute for Radio Astronomy (ASTRON)

Big Data

Projects collecting sensor data over a longer period of time can enter the **petabyte** (=1,000 terabytes) regime. Everything becomes difficult at a petabyte of data storage and beyond. While the capacity of storage units such as hard disks and tape are continuously increasing, in the petabyte regime solutions are much more simple minded: put as many units as you can afford in a row. But this requires complex systems to manage the data, repair errors and find specific data back.

The datasets are too large to move (better to say: **copy**) easily to the computers and hence a new era started in data centers: bring the computers to the data: actually this is the start of the "**data-centric systems**" replacing the classical **compute-centric** paradigm.

At our computing center in Groningen we hosted on the same racks of hardware the data of the **LOFAR radio telescope**, with single files spread over many racks, and a billion mini files for an Artificial Intelligence handwritten text recognition project, next to the medical **Lifelines** data. This created conflicts because of different standards, access patterns and levels of security. The best solution was to separate the domains, recognizing that each domain has its own history and socially developed identity. The lesson learned is that **Big Data** handling has to deal with different sociologies in different domains. In the 2000s a lot of "heal the world" Big Data e-science systems failed. Indeed, finding the balance between common technology and domain-specific approaches is still critical in modern Information Technology. The ambitious European **Open Science** initiative — have all research data publicly available — has to recognize the different domains to avoid these traps. Eventually, this will dominate the work.

Petabyte storage by putting many units in a row. In 2019 16-terabyte disks appeared, and up to 330-terabyte tape cartridges.

Petabyte
Big Data
Open Science
Compute-centric
Data-centric
Being copied
Lifelines
LOFAR

FIRST PICTURE OF A BLACK HOLE

A worldwide network of telescopes collected petabytes of data, which were combined using atomic clocks to form one 10x10-pixel image: the very first picture of a black hole. **Simulations** of black holes are important in both the creation and interpretation of this image.

Credit Katie Bouman, Caltech

One of the test images the EHT teams had to reproduce was this snowman. Being prepared for the unexpected is important in science, to avoid biasing your measurement.

Watching a Black Hole

IN VITRO — *BY HEINO FALCKE*

To create the first image of a **black hole**, we needed a telescope as big as the Earth. Radio telescopes around the world, from Europe to the Americas and from the North to the South pole, worked together as one "Event Horizon Telescope" (EHT). Together they collected 3.5 petabytes of information on a single spot in the heavens with the size of an orange on the Moon. The amount of data was so large, that it could not be transferred through the internet. We had to fly the physical hard disks to two giant supercomputers. There the raw data was reduced from petabytes to mere gigabytes, and finally to the 10x10-pixel black hole image, consisting of a few hundred kilobytes. In the end, we have reduced our data 10 billion times!

Combining data from telescopes around the world is like using one giant telescope which consists mainly of holes, with only a few small areas that are able to collect light. This means we can never measure all parts of the black hole image's structure (for that we would need the entire Earth to be covered in telescopes!). So, we had to reconstruct the black hole image with limited information: gaps in our knowledge of the image's structure. These gaps blur our image, but in a very well-defined way, allowing us to correct for it. To achieve this, it is crucial to know what we don't know: the gaps in our information. There are many algorithms that can use this knowledge to remove the blurring. But which one should we use? If we choose an algorithm with an innate tendency to produce rings, we are just fooling ourselves! So, we organized image reconstruction challenges: contests between different teams, each with their own algorithm. Every team received various sets of simulated data, without knowing in advance what their resulting images should show. We gave them widely different images: a ring, a disk, jets, and even a snowman! The tested algorithms were used to create our first image of a black hole.

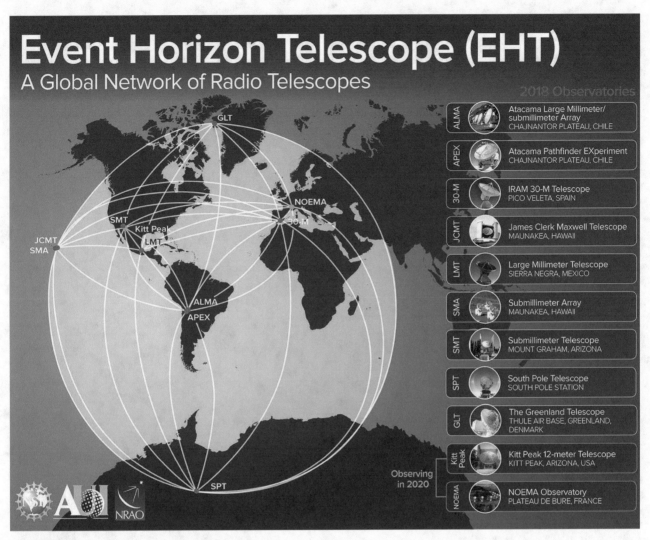

Event Horizon Telescope (EHT)
A Global Network of Radio Telescopes

2018 Observatories

ALMA		Atacama Large Millimeter/submillimeter Array CHAJNANTOR PLATEAU, CHILE
APEX		Atacama Pathfinder EXperiment CHAJNANTOR PLATEAU, CHILE
30-M		IRAM 30-M Telescope PICO VELETA, SPAIN
JCMT		James Clerk Maxwell Telescope MAUNAKEA, HAWAII
LMT		Large Millimeter Telescope SIERRA NEGRA, MEXICO
SMA		Submillimeter Array MAUNAKEA, HAWAII
SMT		Submillimeter Telescope MOUNT GRAHAM, ARIZONA
SPT		South Pole Telescope SOUTH POLE STATION
GLT		The Greenland Telescope THULE AIR BASE, GREENLAND, DENMARK
Kitt Peak		Kitt Peak 12-meter Telescope KITT PEAK, ARIZONA, USA
NOEMA		NOEMA Observatory PLATEAU DE BURE, FRANCE

Observing in 2020

Credit Hotaka Shiokawa, Event Horizon Telescope collaboration

left Black hole simulations, like the one seen here, are very important to test the algorithm that creates our black hole image, but also to interpret the results. We simulated 60,000 black holes with different properties, and used **LOFAR** software to see what these would look like in an actual observation. By comparing these simulated images to the real results, we can figure out the properties of black hole systems (such as their viewing angle and accretion speed) and also test Einstein's theory of **General Relativity**.

Credit Event Horizon Telescope collaboration

above Every telescope in our network measured only a part of the structure of the black hole image. We used atomic clocks to synchronize all these parts, in order to form one single picture. These clocks are so accurate that they will only lose one second every 100 million years. Still, the resolution required to observe a black hole means we are reaching the accuracy limits of even the best atomic clocks.

Black hole
Numerical simulations
LOFAR
General Relativity

Black hole
Lensing
General Relativity
System of Science
Eyes

WATCHING A BLACK HOLE

IN VIVO — BY HEINO FALCKE

What you actually see here is not the very first image of a black hole, but of its shadow. We introduced this term in the year 2000, when we discovered that the shadow is the most robust signature of large black holes, such as this supermassive one at the heart of galaxy M87. No matter from which side we view a black hole, or how fast it is rotating, we will always see this shadow and its surrounding ring of light. This shadow is a signature of the extremely curved spacetime around the black hole, which bends the path of light rays until they disappear into the Event Horizon: the region from which no light can ever escape. Just like an actual shadow, the black hole's shadow is not completely black. We still see a vague glimmer of light, which is emitted by the forward jet in front of the black hole, pointing towards us. The ring of light that we see around the shadow is emitted by the counter jet at the other side of the black hole, pointing away from us. Why do we still see this light? Through the extreme gravity, light is curved all the way around the black hole towards the Earth: an extreme form of gravitational lensing. The fact that we see the full ring means that we observe the jet almost face-on.

Now, why does the bottom of the ring look brighter than the top? The gas forming the accretion disk around the black hole needs to travel with ~50% of the speed of light in order to avoid falling into the Event Horizon. At these relativistic speeds, gas moving towards us appears brighter than gas moving away from us. In other words, this bright blob of light at the bottom reveals that this part of the accretion disk is moving towards us.

In short: light and the way it is curved around the black hole offers us a treasure of information on the features of this black hole system, and also on the nature of space and time themselves. Information transfer is of crucial importance for Einstein's theory of General Relativity. Information, when carried by particles, cannot be transmitted faster than the speed of light, and it is through light that we obtain the vast majority of our information about our Universe. We gather this light through our eyes or telescopes, convert it, interpret it, and obtain information about the world. In this sense information creates our reality, both in daily life and in our System of Science. Einstein, for example, asked himself: "What is time, what is space?" Without measurements, he could never have answered these questions. Thus space and time are defined by how we measure them, through light. In Einstein's theory light, which transfers information, is the only thing which is absolute. Time and space, on the other hand, are relative: they depend on the way we observe them, our speed and our acceleration. In this sense, information even creates space and time.

Credit Event Horizon Telescope collaboration

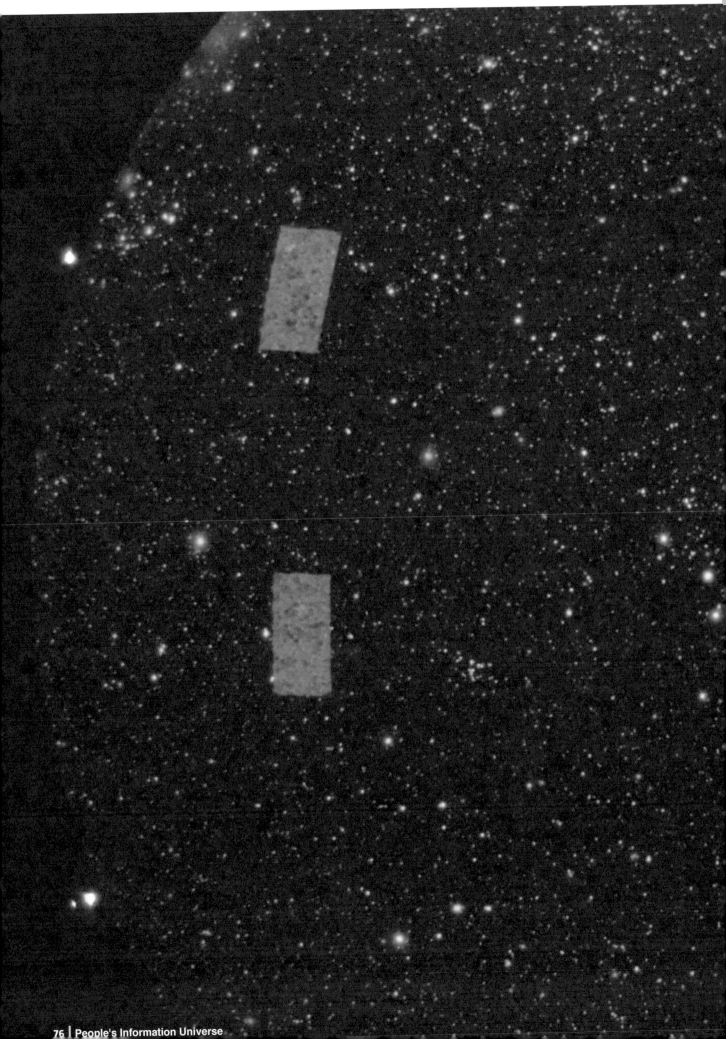

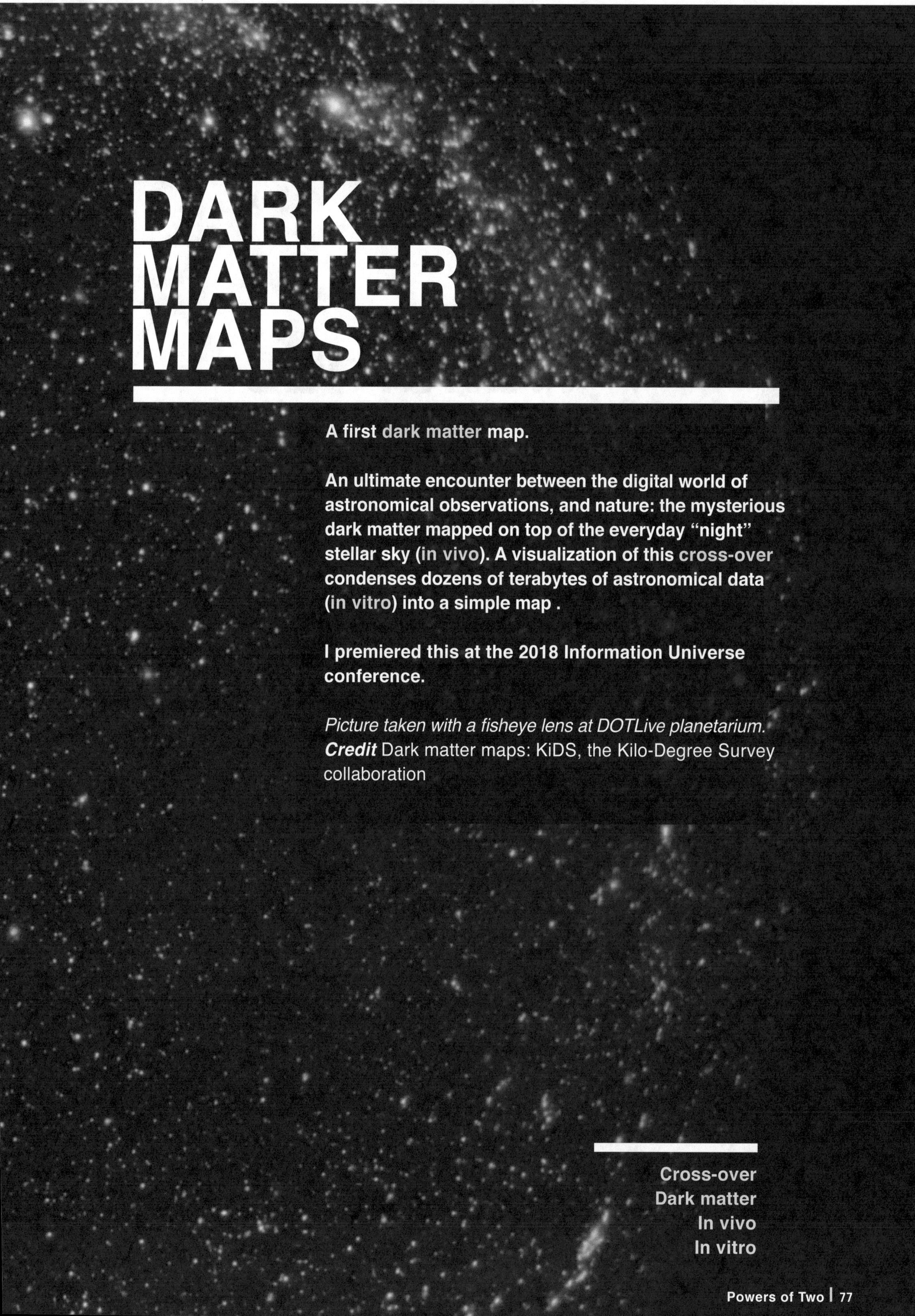

DARK MATTER MAPS

A first dark matter map.

An ultimate encounter between the digital world of astronomical observations, and nature: the mysterious dark matter mapped on top of the everyday "night" stellar sky (in vivo). A visualization of this cross-over condenses dozens of terabytes of astronomical data (in vitro) into a simple map .

I premiered this at the 2018 Information Universe conference.

Picture taken with a fisheye lens at DOTLive planetarium.
Credit Dark matter maps: KiDS, the Kilo-Degree Survey collaboration

Cross-over
Dark matter
In vivo
In vitro

62 bits
$$2^{62} = 5 \times 10^{18}$$
~600 petabytes

Hundreds
of petabytes

POINTERS

With pointers, one can connect everything in the Information Universe. Pointers are often inserted in metadata (data about data) — an ultimate and magic tool for dealing with Big Data. It is possible to create unique pointers to hundreds of petabytes of data, using a string of 64 bits.

With hundreds of petabytes in stored data, the **Big Data** world is really in action. Data sets are often globally distributed and managed by collections of 64-bit numbers called **links or addresses**. It is possible to create unique pointers (links, references or addresses — it's all the same thing) to hundreds of petabytes of data, using a string of less than 64 bits. This is truly a miraculous property of the Information Universe — a 62-bit string of bits, as printed on the top of this page, provides a unique address to any of 600 petabytes. In practice **64-bit** numbers are used as 8 bytes, which is practical technology-wise.

The 64-bit pointer can connect everything in the Information Universe. This is what makes pointers so powerful and indispensable in any form of Big Data, not only for astronomical research, but also for companies like Google, Amazon and Facebook.

Pointers are often inserted in **Metadata** (data about data) — an ultimate tool for dealing with Big Data. For a typical astronomical image about 100 entries are used to describe the image: its origins, the telescope used, size of a pixel, etc. Astronomers were rather early with standardizing these Metadata with the FITS format (1976). Nowadays, most of the efforts in Big Data are in defining the protocols and **standards** for Metadata. There is still an enormous difference in the state of the art for Metadata in different disciplines.

In spite of its crucial importance, there is no physical theory which explains what a pointer is — we are just doing something.

This picture looks like a painted map of the
continents on Earth. It is not, it depicts the links
between friends on Facebook (December 2010).
Note, the absence of China and Russia which
have their own social networks.
Credit Paul Butler, Facebook

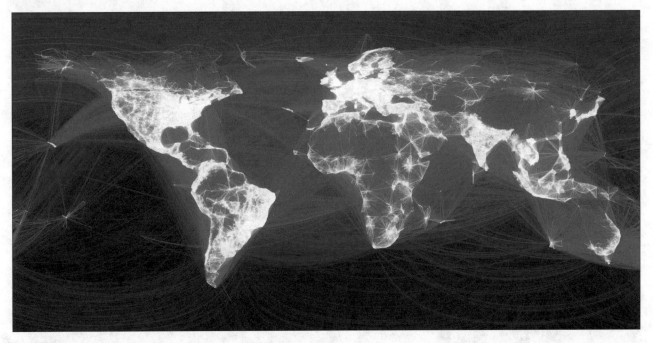

The links between friends of 2 billion Facebook
users in 2017, demonstrating the incredible power
of links in Big Data networks. Petabytes of data are
visualized in a single picture, and by comparing the
two pictures the growth of the network over around
seven years can be easily spotted.
Credit Paul Butler implemented by Jason Sundram

Big Data
Standards
Links — addresses
64-bit
Metadata

BIG DATA TELESCOPES

While current telescopes collect astronomical datasets by the terabytes, the next decade will bring a thrilling zoo of Big Data telescopes: the Rubin telescope with its 3.2-gigapixel sensor array, the Euclid satellite and the Square Kilometer Array radio telescope (SKA) collecting hundreds of petabytes. These enormous amounts of data need a whole new approach to data management. SKA's data rate is equal to the whole internet. For the Euclid satellite this challenge is largely addressed using the specific data-centric approach of the author's research group: downloading the Universe, described on the next pages.

Picture of LSST under construction— Chile
Credit Gianluca Lombardi / LSST/AURA/NSF

Credit ESA

FUTURE — MORE THAN HUNDREDS OF PETABYTES
STAGE IV COSMOLOGY

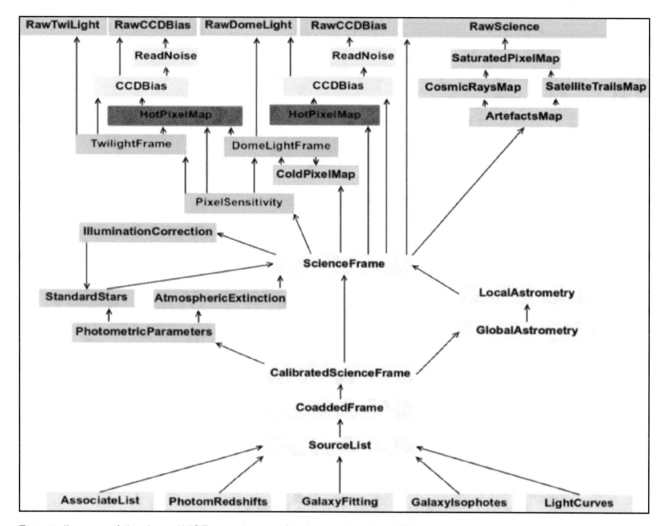

Target diagram of the Astro-WISE system used for processing the VST astronomical surveys.

While in classical **compute-centric systems** computations are modelled in a forward chain, in modern **data-centric systems** it is very beneficial to model the process in a backward chain, recognizing that any result is causally related to a source, whether it is in our computers (**in vitro**) or on the sky (**in vivo**). This technique facilitates the quality control of results, but it is also essential for discriminating between **Facts and Fakes**.

Information is manifest in relation to its environment by **being copied**: some of the bits of the raw astronomical images are copied/printed in this book as images of the sky and eventually they are copied into your brain! The universe can be seen as a spreadsheet, in the way we map all the **links** with **64-bit** pointers on our computers (in vitro), but also in nature (in vivo).

Perceiving the Universe as a spreadsheet links bit to It.

Compute-centric
Data-centric
Links — addresses
Being copied
Astro-WISE
Target
In vivo

In vitro
Facts and Fakes
64-bit
Petabyte
Metadata
Fisher information

THE UNIVERSE AS A SPREADSHEET

DOWNLOADING THE UNIVERSE

The big astronomical surveys record huge datasets with the sensor arrays at the telescopes. The events at the detectors are actually quantum events interfacing the real physical world —Information Universe (IU) **in vivo** — to the bits of information that are endlessly **copied** in our electronics and computers of the IU **in vitro**. The raw data have to be processed using a zoo of computer programmes with around a million lines of code, to remove the fingerprints of the instrument, telescope and the effects of the Earth's atmosphere, which all change in time during the years of observations of a survey. But also the insight of the programmers into these variations change over time, better methods are developed and coding errors are made.

In the **Astro-WISE** information system we organize this evolving Big Data process by carefully mapping the causality between all data products and the sea of raw data taken at the telescope depicted at the top of the **Target** diagram on the left page. The causality is mapped as dependencies/links (arrows) to other intermediate data products up to the raw image data. The object-oriented programming technique allows us to implement this in a straightforward way. In fact, the whole processing of **petabytes** of survey data is organized by maintaining all the links between the data items, and storing these in **metadata** of the images in a relational database, while the images themselves could be anywhere on the globe. Since everything is changing in time this is a very beneficial approach which allows the user to ask any relevant questions, particularly regarding the quality of the result — each data product "knows" all its dependencies, and even when the product has not been made or when its dependencies are out of date, the product is computed on the fly. This approach is designed to be identical to the causality in the physical world (IU **in vivo**) at the other side of the quantum event at the detector. There is suspicion that there is no fundamental difference between the two Universes — hence the overall view of the Universe as a spread sheet, linking It to bit, or Frieden's **I** to **J** in "Science from Fisher Information"[10) 11).

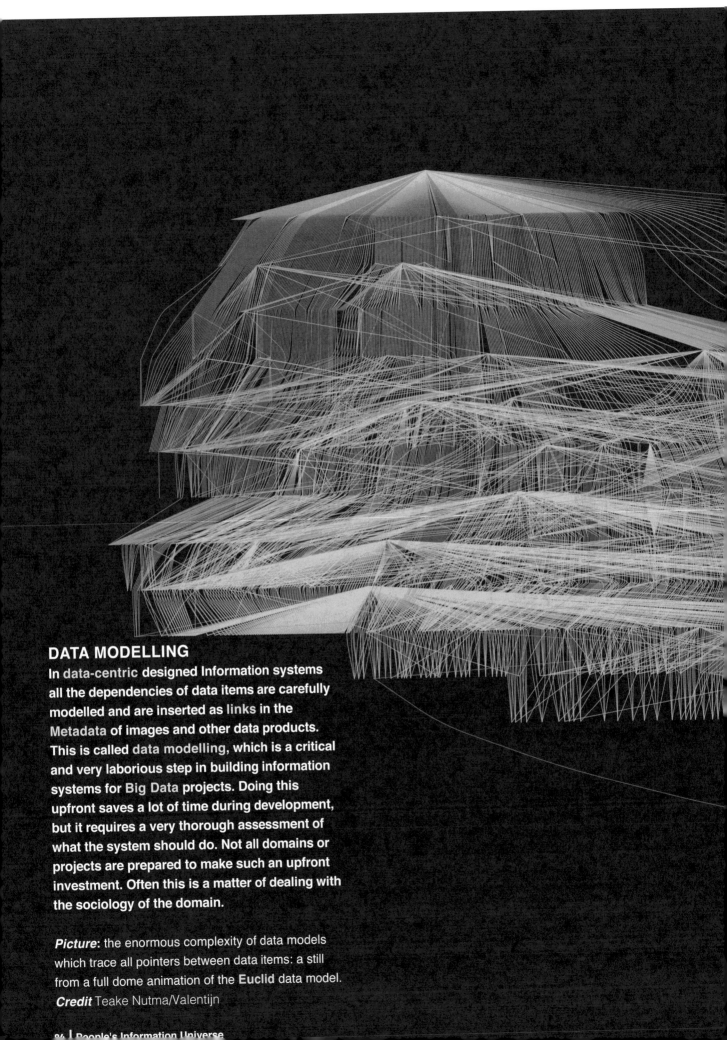

DATA MODELLING

In data-centric designed Information systems
all the dependencies of data items are carefully
modelled and are inserted as links in the
Metadata of images and other data products.
This is called data modelling, which is a critical
and very laborious step in building information
systems for Big Data projects. Doing this
upfront saves a lot of time during development,
but it requires a very thorough assessment of
what the system should do. Not all domains or
projects are prepared to make such an upfront
investment. Often this is a matter of dealing with
the sociology of the domain.

Picture: the enormous complexity of data models
which trace all pointers between data items: a still
from a full dome animation of the **Euclid** data model.
Credit Teake Nutma/Valentijn

METADATA
DATA-CENTRIC APPROACH

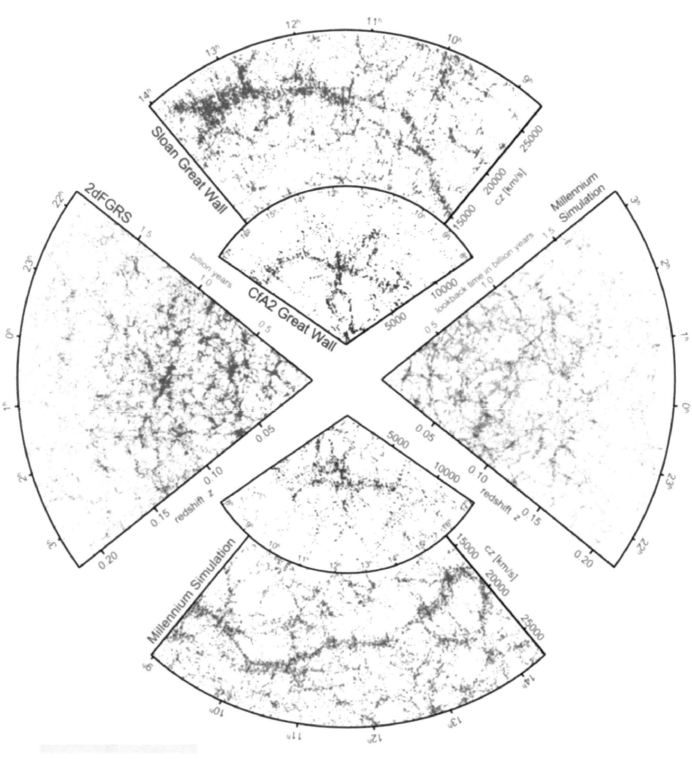

Numerical simulations
Dark matter
Dark energy
ΛCDM
Cross-over
In vivo
In vitro

THE INFORMATION UNIVERSE *IN VIVO – IN VITRO*

For me this is an epic picture of the Information Universe, published by Volker Springel et al. in 2006[12]. It represents the large-scale distribution of galaxies around us, with the Earth at the centre of the picture, showing increasing distances up to about 1.3 billion light-years at the outer ring. The elongated structures are galaxies along filaments spanning hundreds of millions of light-years. The regions in between seem to be rather empty (voids).

The pronounced large-scale structure of our Universe is stunning, but even more amazing is that the points in the blue and dark blue cones at the left and top are determined from direct observations of the distances (redshifts) of true galaxies, while the red and brown cones at the right and bottom result from the Millennium supercomputer numerical simulation of the growth and evolution of galaxies, from their creation in the computer code onwards. The typical filamentary structures emerging in the computer simulations are remarkably similar to the look and feel of the truly observed galaxies. The computer code in the simulations assume the ΛCDM model of our Universe, with Cold Dark Matter and dark energy which accelerates the expansion of the Universe. Allthough the simulations predict far too many dwarf galaxies, the picture highlights the success of the ΛCDM model describing our Universe, but it also amazingly features the cross-over of the in vivo (observed) and in vitro (simulated) Information Universe.
Credit Volker Springer 'NATURE'

After 25 years an Indian boy found his natural mother (in vivo) by searching on Google maps (in vitro).

Bada

Railwa

Gngur ghat

Bhimkund Mandir
भीमकुंड मंदिर

Google

IN VIVO – IN VITRO
STORY OF THE LOST BOY

Information is timeless and knows no boundaries. It crosses over the in vivo and the in vitro Information Universe. This concept is well illustrated through daily life stories involving time. There are many — but this is one of my favourites. In 1986, the five-year-old Saroo lost sight of his older brother and fell asleep on a train in India for 14 hours, all the way to Calcutta. Since he did not know the name of his hometown, he had no alternative than begging on the streets. After a year he was taken up by an orphanage which found him an adoptive family in the very far Tasmania. Twenty-four years later he decided to search for his birthplace: he knew no name, only that it was ~14 hours by train from Calcutta. He scrutinised Google Maps within a radius of 1,200 kilometres from Calcutta. Incredibly, he recognized a waterfall and the area around the river and could navigate to his home place. A year later he decided to go there, where he found his mother who had been waiting for him all those years — but who also had to tell him that his brother was found dead at the railway station where Saroo had lost sight of him.

Credit Saroo Brierley
Credit Google Maps, DigitalGlobe Map data (2018)

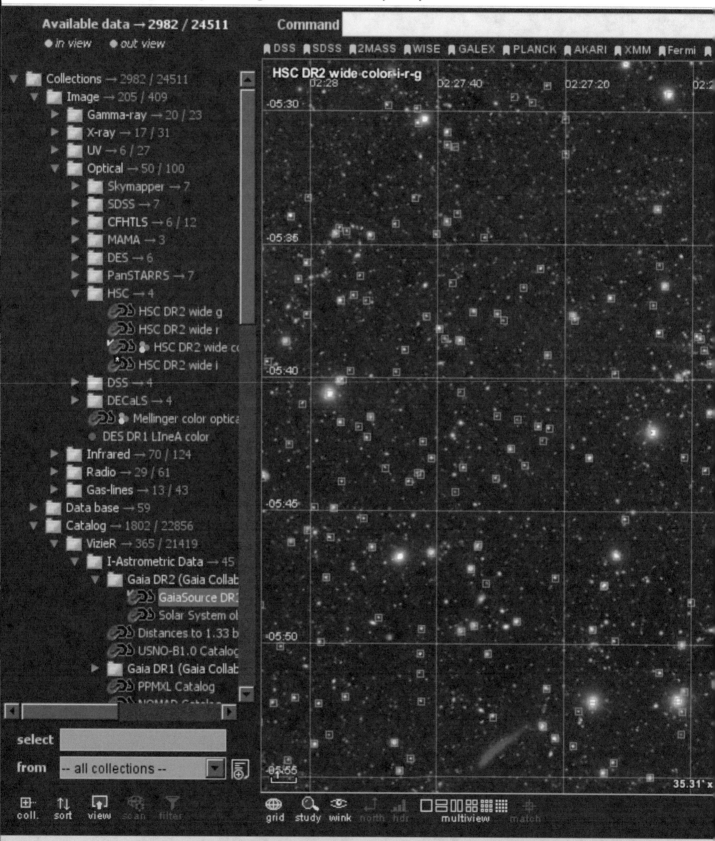

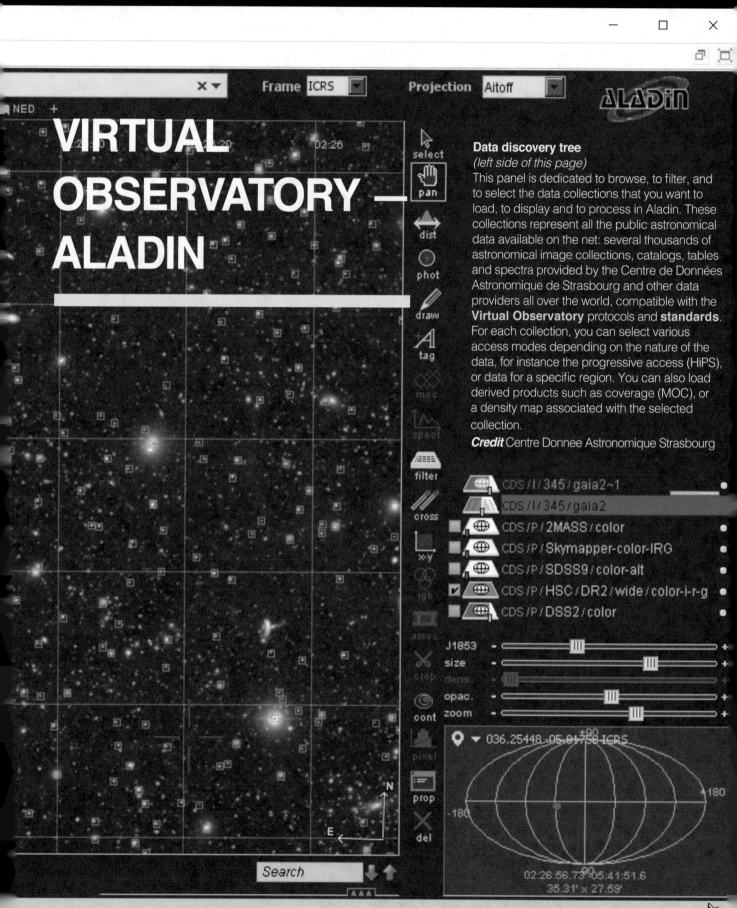

VIRTUAL OBSERVATORY — ALADIN

Data discovery tree
(left side of this page)
This panel is dedicated to browse, to filter, and to select the data collections that you want to load, to display and to process in Aladin. These collections represent all the public astronomical data available on the net: several thousands of astronomical image collections, catalogs, tables and spectra provided by the Centre de Données Astronomique de Strasbourg and other data providers all over the world, compatible with the **Virtual Observatory** protocols and **standards**. For each collection, you can select various access modes depending on the nature of the data, for instance the progressive access (HiPS), or data for a specific region. You can also load derived products such as coverage (MOC), or a density map associated with the selected collection.

Credit Centre Donnee Astronomique Strasbourg

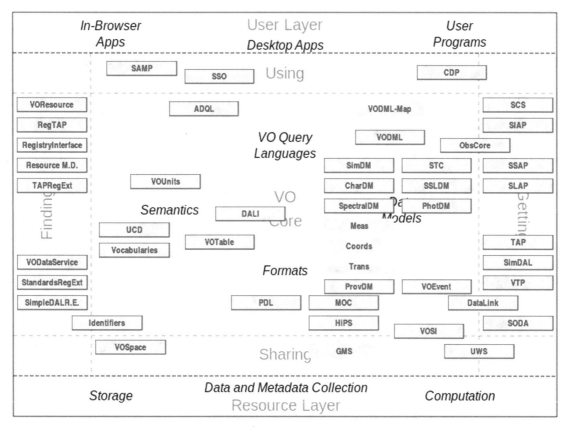

Users   **Computers**

In-Browser Apps   User Layer   User Programs

Desktop Apps

SAMP · SSO · Using · CDP

VOResource · ADQL · VODML-Map · SCS

RegTAP · SIAP

RegistryInterface · VODML · ObsCore

Resource M.D. · SimDM · STC · SSAP

TAPRegExt · VOUnits · CharDM · SSLDM · SLAP

VO Query Languages

Semantics · DALI · SpectralDM · PhotDM · Data Models

UCD · VOTable · Meas · TAP

Vocabularies · Coords · SimDAL

VODataService · Trans · VTP

StandardsRegExt · Formats · ProvDM · VOEvent

SimpleDALR.E. · PDL · MOC · DataLink

Identifiers · HiPS · SODA

VOSpace · VOSI · SODA

Sharing · GMS · UWS

Storage   Data and Metadata Collection   Computation

Resource Layer

Registry · **Data Access Protocols**

Providers

Previous page: just a view of the Universe
Credit Centre Donnee Astronomique Strasbourg

I made this screenshot on my computer using the Aladin app[13] which is built and freely distributed by the International **Virtual Observatory** Alliance (IVOA). Development of Aladin started as early as 1999 at the Centre Donnée Astronomique Strasbourg, and was and still is ahead of its time. By now, it provides a visual interface to thousands of image collections (atlases) from astronomical surveys all over the world, and tens of thousands of catalogues with spectra and data on astronomical objects. I loaded images taken with the Japanese Subaru telescope and its Hyper

Suprime-camera at the summit of the Mauna Kea volcano in Hawaii and I tuned the field of view to correspond more or less to the diameter of the full Moon. These wide field images give a fantastic wide and deep view of the universe. Then I asked Aladin to identify the stars observed by the Gaia space telescope and marked them with a blue box. I could equally well have loaded data from X-ray telescopes, infrared, radio or gamma ray surveys. This is truly amazing, although my students now take it for granted. But note, the second Gaia data release contains ~1.3 billion stars in our own Galaxy. Beyond these stars identified by Gaia, most of the fainter spots are galaxies, each containing billions of stars. It is estimated there are about 100 billion

galaxies in our Universe and around 5 billion can be detected by Hyper Suprime-cam and the **Euclid** satellite. The **in vivo** – **in vitro** Information Universe at large at your finger tips.
Credit IVOA

Virtual Observatory
Euclid
In vivo
In vitro
Google Maps
World Wide Telescope
Open Science
FAIR
Standards
Astro-WISE
Links — addresses
ASCII
Metadata

VIRTUAL OBSERVATORY *IVOA*

Wouldn't it be great to be able to navigate on your screen through the sky in the same way we use **Google Maps** and Google Earth? Well, it is available in Google Sky[14] and **World Wide Telescope** originally developed by Microsoft[15]. For professionals there is the Aladin app developed by the International **Virtual Observatory** Alliance (IVOA), a world-wide organization with the vision that astronomical datasets and other resources should work as a seamless whole: the Information Universe in vitro. Many projects and data centres worldwide have been working toward this goal for a long time. I read in one of our first proposals for the Virtual Observatories in 2001: *"The AVO can be defined as a science facility in which the "telescope" is represented by the network of existing Archives, while the "instrument" is the suite of software tools that query and analyse the data, without physically moving them from the Archives, and responding to specific scientific questions that are posed by the "observer".* And so it happened: next to the Aladin tool a large amount of services for data sharing and analysis have been developed, which are used world wide. Today's trend towards **Open Science**, with all research data made public and accessible according to the **FAIR** principles, is very close to the ideas of the **VO**, and actually also my own **Astro-WISE** information system; both have been developed by astronomers for decades. **FAIR** stands for data items being **F**indable, **A**ccessible, **I**nteroperable and **R**eusable.

What are the lessons learned from this? Above all: the enormous effort it takes to define and update all the **standards** to achieve inter-operability. The graphs on the left page depicts the most common standards developed by the **VO**. Most of them define **links** — like **ASCII** — to define **Metadata** and locations of data repositories, but also protocols to access the data. These standards are discussed, documented, implemented, tested, qualified and then preached to the community. Many of them have many versions during the decades of development. All this is pretty specific for the astronomy domain. Other domains have a different sociology: work cultures, protocols, and ways of managing projects. I frequently advise science policy makers not to underestimate what it takes to achieve truly **open science** for all the different domains. Even with all this effort, the experiment-specific processing can not be handled, as too much experiment-specific knowledge is required. Systems like Astro-WISE, take care of all the calibrations of observational data, and work as a front-end for the Virtual Observatory.

Mark Allen, chair IVOA, at his closing address at the 40th interoperational meeting of the IVOA, at DOTlive-planetarium, Groningen. October 2019

50 qubits
2^{50} states

ATOMS AND ELECTRONS

In the quantum world of atoms and electrons, 50 quantum bits or **qubits** can assume 2^{50} values all at the same time. And 500 qubits can take on 2^{500} values in parallel, a number larger than the number of atoms in the universe! How is this possible? And what does this bring us?

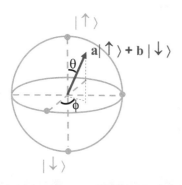

Bloch sphere illustrating **Qubits**.
Electron spin as quantum bit.
Spin pointing up ⟶ bit value "0"
Spin pointing down ⟶ "1"
Spin pointing in other direction ⟶ bit value simultaneously "0" and "1", with specific weights.
Credit QUTech, TU Delft

Qubits
BY LIEVEN VANDERSYPEN

Think of the spin of a single electron as the smallest imaginable magnet. Like a ship's compass needle, it can point in any direction, and its state can be described by an azimuthal angle φ and a polar angle θ.

Naturally, for every additional compass needle, I simply need another two angles to describe their state. Spins being quantum mechanical objects, the situation is completely different than for classical compass needles: for every spin I add, the **complexity** of the system *doubles*, i.e. we need twice as many real numbers (such as angles) to describe the system state. Like any exponential, this quickly goes through the roof. The state of just a few hundred spins contains more variables than there are atoms in the known universe!

This exponential complexity arises directly from the quantum mechanical nature of the spins. It is also intimately connected to the concept of **entanglement** — see page 99.

Now let's get practical. If I use electron spins as **quantum bits** in a computer (I actually do exactly that in my research) the bit string can assume exponentially many values at the same time, and I can compute on all these values in parallel. Taking advantage of this exponential power and complexity, so-called **quantum computers** are capable of solving important problems in physics, chemistry, materials science and mathematics that are simply beyond the reach of conventional supercomputers. These problems in turn find broad application for addressing major societal challenges, from health to energy, climate and security. If only we had a quantum computer. Read on about where we stand on the next page!

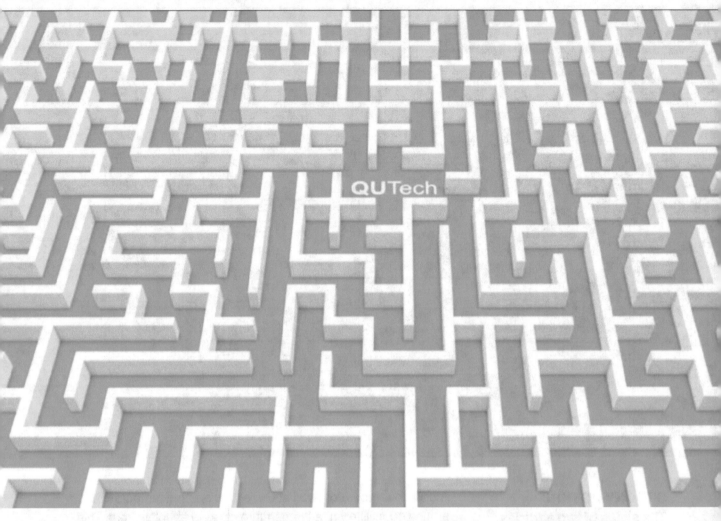

In the classical world, there is no better strategy for escaping from a maze than trying one possible path after another. A quantum computer can find the way out of maze by exploring all possible paths simultaneously.
Credit QuTech, TU Delft

01 & 10

This combination of values of two quantum bits is very special, as it corresponds to an **entangled** state. Measurement of either qubit gives a random outcome, either 0 or 1. Subsequent measurement of the other **qubit**, will necessarily give the opposite outcome as the first measurement. The state of the two qubits is linked by quantum entanglement, which has no classical equivalent.

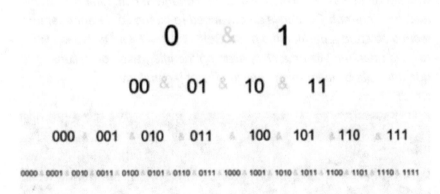

0 & 1

00 & 01 & 10 & 11

000 & 001 & 010 & 011 & 100 & 101 & 110 & 111

0000 & 0001 & 0010 & 0011 & 0100 & 0101 & 0110 & 0111 & 1000 & 1001 & 1010 & 1011 & 1100 & 1101 & 1110 & 1111

For every *classical* bit that we add to a bit string, the number of possible values doubles — the Powers of Two. For every **quantum bit** that we add to a string of quantum bits, the number of possible values that can exist simultaneously doubles. The exponential complexity of such "quantum superposition states" lies at the heart of the power of quantum computers.
Credit QuTech, TU Delft

Entanglement
Qubits — quantum bits
Quantum computers
Complexity

SPOOKY ACTION

The states of two particles can be intimately linked (entangled), no matter how far they are separated. What Einstein famously dismissed as "spooky action at a distance", can now be established on demand at TU Delft in the Netherlands. Prof Vandersypen explains how researchers from his group and institute both create and apply this entanglement in the laboratory.

Entanglement
Qubits
Quantum computers
Complexity

A dilution refrigerator used at QuTech to reach a temperature around 10 millikelvin, a temperature close to absolute zero. This extremely low temperature is reached through successive stages of cooling and are needed to minimize the thermal fluctuations that otherwise would perturb the fragile state of the **quantum bits**.
Credit KLAPSTUK for QuTech

At QuTech in Delft, 250 scientists, engineers and support staff are working on scalable prototypes of both a quantum computer and on a quantum internet. QuTech is an advanced research center for quantum technology and a partnership of TU Delft and TNO. Our approach is to build quantum bits on a chip, allowing the integration of millions of quantum bits by leveraging semiconductor technology.
Credit Qutech, TU Delft

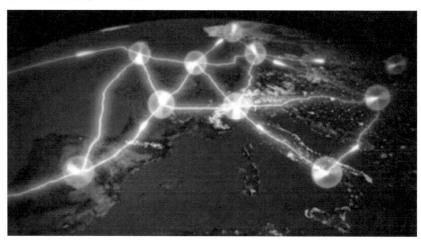

QUANTUM COMPUTERS GET REAL

BY LIEVEN VANDERSYPEN

The realization of a **quantum computer** offers the wonderful promise of solving important, otherwise insurmountable problems. But it also presents a formidable challenge: the state of "**quantum bits**" is notoriously fragile, and is easily disturbed by the slightest microscopic fluctuation in their vicinity. In order to minimize these fluctuations, quantum bits are cooled down to close to absolutely zero temperature. The image on the left page shows the inside of a special refrigerator capable of reaching 10 millikelvin and operated in vacuum.

Despite these challenges, in the Fall of 2019, a team from Google achieved control of 53 quantum bits in a tour-de-force experiment. This particular device was only able to compute itself: it was fed with a random set of instructions, yielding a random outcome, but the **complexity** of computing which random outcome was beyond the abilities of classical computers. Today, much work is focused not only on increasing the number of qubits that can be controlled, but also on understanding how to solve real-world problems with the hundreds of quantum bits that are expected to become available in the next few years. Meanwhile, at QuTech we have already

made real quantum bits available 24/7 to anyone in the world interested to explore quantum computing for themselves. This is done via a publicly accessible online platform called Quantum Inspire (**https://www.quantum-inspire.com**).

Quantum Inspire is the first European quantum computer in the cloud, and the first platform with two different types of qubits under the hood: semiconductor spin qubits and superconducting qubits. Unleashing the full power of quantum computers, however, will likely take millions of quantum bits. Only then the system contains sufficient redundancy for recovering from the unavoidable errors introduced by the fluctuations from the qubits' environment. Achieving such a machine poses one of the biggest and most interesting scientific and technological challenges of this day.

In the farther future, we imagine quantum computers to be connected together via quantum links, to form an inherently secure quantum internet. An important milestone in this direction was an experiment by Ronald Hanson and colleagues, who demonstrated quantum entanglement over 1.3 km across the TU Delft campus and closed for the first time all the loopholes that remained in earlier experiments.

May 20, 2002

September 2, 2002

October 28, 2002

December 17, 2002

February 8, 2004

above V838 Monocerotis imaged by Hubble ACS. The interstellar dust is illuminated by the red supergiant star in the middle of each image, which gave off a flashbulb-like pulse of light. When comparing the images from different dates the dust appears to move with ~12 times the speed of light. It's an illusion: the particles do not move, but the orientation of the beam from the star moves faster than light over the dust shell. The motions of particles are constrained by the speed of light — information, such as the track of the beam hitting the medium, can be faster.
Credit NASA.

below Verlinde postulates that the **dark energy** in the universe is entangled over regions of space that could never have communicated with each other, and that as a result the motions of massive objects imitate the effect of dark matter. This continuous influence from dark energy is non-local and connected instantaneously, in the same way we continuously feel the gravity of the Sun. However, changes in the effect of **dark energy** will be transmitted at a finite speed. Just like changes

in gravity, such as the motion of the Sun, would be transmitted (according to Einstein's theory) by **gravitational waves**, which were recently observationally confirmed to travel at the speed of light.
Credit NASA, ESA, H.E. Bond (STScI) and The Hubble Heritage Team (STScI/AURA)

In entangled particles information in **Qubits** *is shared at a distance instantaneously.* **Credit** *Doug Cohen, Louis Tarantino, from 'Quantum Entanglement Simplified'*

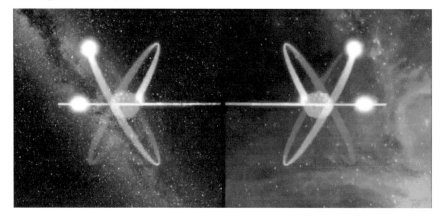

ENTANGLEMENT

In quantum physics two particles can be connected and have **entangled** states when they form a composite system with an overall state that can be expressed as a sum of products of the states of the two particles.

In practice, this means that changing something in one particle will change the result of a measurement of the other particle. This happens *instantaneously*, even when the particles are light-years separated and the speed of light would take years to send a message across. This is called **non-locality**, and apparently particles can be connected across the universe unconstrained by spacetime and the speed of light. This has been experimentally verified[16], and even **teleportation** of particles based on this principle has been achieved[17]. In this form information is not constrained by the speed of light, but non-locality cannot be used to actively send messages from A to B. Quantum computer experts would love this: use **Qubits** to build an interstellar infinitely fast internet based on non-locality. But this is impossible, as you always have to transmit with classical means the outcome of the measurement at A to B to decode the signal at B.

Non-locality is also referred to as the EPR paradox after Einstein, Podolsky and Rosen who thought that this prediction of quantum mechanics should disqualify the theory[18]. In fact,

Bell's work[19] and its experimental verification[20] proved that non-locality is true in quantum mechanics. Non-locality demonstrates that information also has its own elementary properties unrelated to spacetime and different from particles, providing a building block of the **in vivo** Information Universe. In the **in vitro** Information Universe, in our computers, a different entanglement is created by humans in the form of **64-bit** pointers. Pointers, **links**, addresses, joins, URLs, associations, annotations and blockchain all connect two states — they all do the same thing — though there is no physical theory explaining how this works.

The beam of a superfast lighthouse can move faster than light over the horizon, simply because there is no particle moving — only information. For example, astronomers observe cosmic light echoes where the overall structure evolves faster than light (*see left page*).

Dark energy
Gravitational waves
Entanglement
Non-locality
Qubits
In vivo
In vitro
64-bit
Links — addresses
Teleportation

THE SQUARE KILOMETRE ARRAY

The Square Kilometre Array will collect data at a higher rate than the current global internet traffic in its endeavour to answer questions about the origin and evolution of the Universe, and its search for extraterrestrial life. Eleven countries join forces to build a radio telescope with a combined surface area of 1 square kilometre.
Credit ICRAR/Curtin

The Australian section of the SKA will contain 130,000 antennas, together collecting 250 terabytes of data every second.

Exabyte — SKA
BY MICHIEL VAN HAARLEM

Everybody knows that astronomers love big telescopes, giving them names like "Very Large Telescope" and "Extremely Large Telescope"; the latter having a light collecting area of a whopping 978 square meters. Radio astronomers like to use even bigger telescopes, like **LOFAR** with a combined collecting area of 100,000 square meters at the lowest frequencies. But this pales in comparison to its successor: the Square Kilometre Array (**SKA**), a radio telescope that will have a combined collecting area of a million square meters: 1 square kilometre. Actually, the SKA is a combination of two telescopes: one on the remote plains of South Africa, and the other in the arid deserts of western Australia. These big telescopes will collect **Big Data**: even more than the current global internet traffic. Every second the ~200 South-African dishes produce 1.1 terabytes, while the 130,000 Australian antennas produce 0.25 petabytes. This results in a grand total of 8 zettabytes per year: eight times the information content of all global data. Storing this amount of information is neither feasible nor affordable, which means the data needs to be reduced in real-time at the two locations. This first reduction step combines the data from all receivers at each site into a mere 5 terabits per second, resulting in a total data-rate of 40 exabytes per year; still far too much to store. Through thousands of kilometres of fibre optic cables, this data is transported to Perth and Cape Town for the second step: Science Data Processing. The raw data is transformed into 3D maps of the sky, which contain information on all the astronomical sources observed by the SKA. This step reduces the data ~70 times, which leaves a manageable 0.6 exabytes per year for astronomers to study. As coordinator of AENEAS, I led a European project to design the network of science data centres that will be used to distribute the final data to astronomers around the world. Having these science centres ready on time is critical to getting the eagerly awaited scientific results, such as tests of Einstein's theory of **General Relativity** and studying the signals from the first stars and galaxies that formed after the **Big Bang**.

Credit ICRAR/Curtin

Together with **Euclid**, Rubin and other telescopes, the **SKA** is part of the Stage IV suite of cosmology projects that will collaborate to unveil the most pressing mysteries of our Universe. Radio astronomers have written more than 2,000 pages filled with research ideas: from studying the evolution of galaxies, **dark matter** and **dark energy**, to searching for the building blocks of life, habitable planets and even Extra-Terrestrial Intelligence (SETI). The scientific research made possible by this telescope is too abundant to describe. What inspires me most is that the SKA will allow us to make the first observation of the Cosmic Dark Ages, the period before stars and galaxies had formed. Although this epoch is the seed of all cosmic structure — the starting point of every cosmological **simulation** — we have no visual

information whatsoever about this time in cosmic history. Visible light simply did not exist yet! But because the SKA observes light at radio frequencies, we can make the first map of the temperature and density fluctuations in this primordial gas, when it was still in the process of forming the first stars and galaxies.

Another inspiring possibility is using groups of pulsars to detect extremely long **gravitational waves**. Pulsars are neutron stars whose steady rotation can be used as very accurate cosmic clocks. Finding a similar anomaly in the ticks of several of these clocks, which are located thousands of lightyears apart, can indicate giant waves in spacetime itself. These are very different from the short gravitational waves observed by ground-based experiments like LIGO and Virgo.

They are only caused by collisions of supermassive **black holes** at the centres of galaxies, or even more violent events like the cosmic inflation period. It's beautiful that, instead of building an experiment, we can use these pulsars as giant gravitational wave detectors that are given to us by the Universe.

LOFAR
SKA
Big Data
General Relativity
Big Bang
Euclid
Gravitational waves
Black hole
Dark matter
Dark energy
Numerical simulations

ENCRYPTION

Encrypted messages should not be decodable by adversaries, be they criminals or hostile countries. **Cryptography** enables secure communications and is one of the few applications which require 128-bit numbers.

FINDING LOGARITHMS

In order to find the logarithm of a number *h* to the **base** *g*, one needs to solve for *x* the following equation:

$$g^x = h$$

Finding logarithms over the set of real numbers is easy, but not over certain cyclic groups. Many cryptosystems in use today need the hardness assumption of the discrete logarithm problem including Diffie-Hellman-Merkle key exchange and the Elliptic Curve Digital Signature Algorithm (ECDSA). They can all be broken by Shor's quantum algorithm.

Cryptography

BY SIMONA SAMARDJISKA

You will hardly ever hear anyone using the number 2^{128} to express a time period or the number of steps to perform a task. Anyone but a cryptographer. Nowadays, a cryptographer would tell you that a cryptosystem is secure if an attacker needs to perform at least 2^{128} operations to break it. What does that exactly mean, and why is it relevant for us? Well, to begin with, cryptographic schemes are the core building blocks of various systems that secure our digital assets and enable secure communication. Cryptography makes breaking the security of these systems computationally as hard as opening a door by trying 2^{128} different keys, or as solving a very hard mathematical problem that requires 2^{128} operations to solve. You would say, of course the attacker can use a computer, or even a super computer or many of them. Wouldn't breaking these systems be a walk in the park then? Turns out, no. An attacker with the processing power of the top 500 supercomputers in the world in 2019 would still need more time than the age of the universe.

Things change, however, if the attacker has at hand a reasonably powerful quantum computer implementing a few thousand logical **qubits**. Such quantum computers don't exist yet — the technology of quantum computers is in its infancy — but it is quite plausible that in a decade or two, we will have powerful enough quantum computers to break cryptosystems considered secure today. The reason is that there exist quantum algorithms that can solve in polynomial time mathematical problems considered hard today. Such is the integer factorization problem which underlies the RSA cryptosystem — one of the most widely used public key cryptosystems today. Using Shor's quantum algorithm[21] invented in 1994, a 3072-bit RSA composite number can be factored in a few seconds. On a classical computer — you guessed right — one needs to perform 2^{128} operations.

Luckily, the arrival of powerful quantum computers will not be the doomsday of our digital security. Many cryptographers (including myself) are working on cryptographic designs for the quantum era. A special branch called post-quantum cryptography deals with designing classical cryptosystems that will be hard even for attackers with quantum computers, and would still require 2^{128} quantum operations to break them.

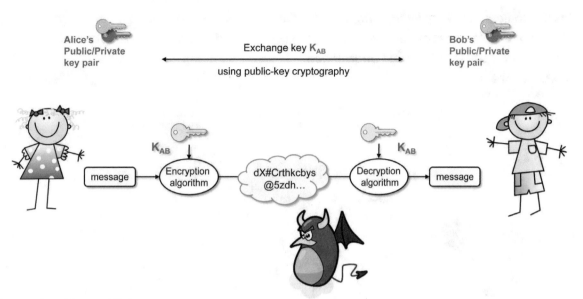

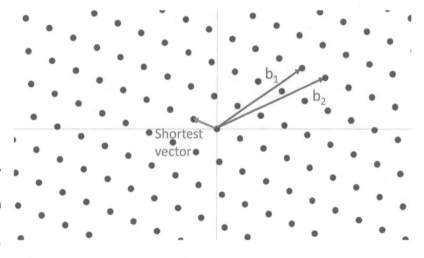

top Two parties, Alice and Bob can communicate secretly by encrypting their messages using a shared 128 bits-long secret key K_{AB} and a secure symmetric key algorithm such as AES. This requires the key to be known by Alice and Bob before they encrypt their messages. They could meet in person to agree on their key, but most of the time this is not an option. Instead, they can use a public key algorithm to exchange the key. In such an algorithm one or each party has a pair of public and private keys. Alice can freely share her public key on the internet, and anyone can use it, for example, to encrypt a message intended for Alice. Only Alice, using her corresponding private key, can decrypt the message. The invention of public key cryptography by Diffie, Hellman[22] and Merkle[23] in the 70s is considered the most important milestone of modern cryptography. Today, public key cryptography has countless applications — it is used to exchange symmetric keys, to digitally sign documents, to authenticate servers, to securely share secrets, to anonymously browse the internet and a whole lot more.
Credit Simona Samardjiska with added elements from pixabay

above Finding the shortest vector in an n-dimensional lattice is one of the fundamental hard problems in lattice based cryptography — a special area of post-quantum cryptography. This problem is known to be among the computationally hardest mathematical problems and it is widely believed that even an adversary with a **quantum computer** cannot solve it in polynomial time. Today there are cryptosystems that are computationally as hard to break as solving the shortest vector problem for large dimensions. The figure depicts the problem in two dimensions. The red dots represent a lattice in a Euclidean space, and b_1 and b_2 are the basis vectors that generate the lattice.
Credit Simona Samardjiska

Qubits
Quantum computers
Cryptography
Base

"What lies at the heart of every living thing is not a fire, not warm breath, not a 'spark of life.' It is information, words, instructions... If you want to understand life, don't think about vibrant, throbbing gels and oozes, think about information."
Richard Dawkins in 'The Blind Watchmaker'

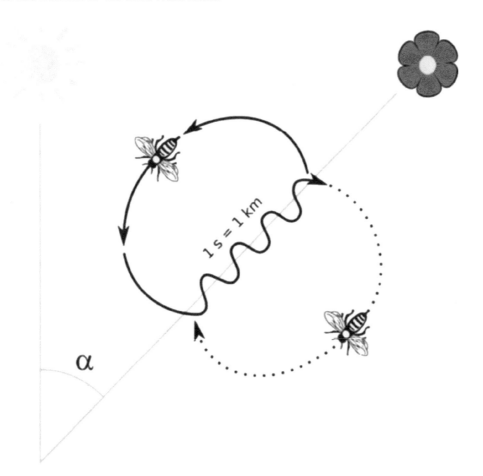

Before the dance the food could be anywhere along a distance of 14 kilometres with uniform probability. From **Shannon** we get:

$$I_{before} = -K \sum_{i=1}^{14000} \frac{1}{14000} \ln \frac{1}{14000} = 13.8 \ bits$$

If we know that the food is located at a distance of 1716 meters, with an uncertainty of 147.2 meters, then assuming a normal distribution we have:

$$I_{after} = -K \sum_{i=1}^{14000} \frac{1}{14000} \frac{1}{\sqrt{2\pi\sigma^2}} e^{-\frac{(i-\mu)^2}{2\sigma^2}} \ln \frac{1}{14000} \frac{1}{\sqrt{2\pi\sigma^2}} e^{-\frac{(i-\mu)^2}{2\sigma^2}} = 9.24 \ bits$$

We can do the same for the direction. All in all 7 bits are communicated by the little worker bee.
Credit Emmanuel Boutet / Courtesy:https://en.wikipedia.org/wiki/Bee

NATURE PROCESSES INFORMATION

BY PETER SLOOT

When Claude Shannon was looking for a name for the equation with which he could calculate the *missing information* in a communication channel[24], it was John Von Neumann who suggested calling it **entropy** saying: "In the first place your mathematical development looks a lot like Boltzmann's. Secondly no one understands entropy so in any discussion you will always be in a position of advantage!" This little anecdote shows that we have no clue what information or entropy is, we only know what it does: *Information reduces uncertainty* which we can measure by the amount of **entropy** change.

Consider a lovely little worker bee just returning to the hive after discovering a nice food source. With a little waggle dance the worker will communicate the location of the food source in terms of distance from the beehive, angle with respect to the position of the Sun, and even how much food her fellow bees can expect there[25]. Using **Shannon's** equation, one can show that the uncertainty in the food location before the dance is about 24 bits and after the dance this is 17 bits, so the bee has communicated 7 bits of information, beautiful[26]!

We can extend this notion of Information processing all the way from the molecules of a cell, to cells in an organism or individuals in a society. These are all nodes in an interwoven information network, collectively transmitting and receiving, coding and decoding information[27].

Where does all this information come from?

Thanks to the seminal work of Maxwell, Boltzmann and Gibbs we now know that the free energy that the Sun pumps into the biosphere of the Earth can be seen as the source of order and **complexity** of living systems. One way to look at this is that part of this Gibbs free energy drives the molecular machinery and builds up the statistically unlikely thermodynamic information structures keeping the organism away from thermodynamic equilibrium! These out-of-equilibrium improbable dynamical structures gain increasing complexity by interaction with their environment through natural selection.

Shannon
Entropy
Complexity
Life

Theoretical physics has not progressed much in the last decennia — some call it a crisis. Likely, a break-through in our own laboratories is out of reach: the highest man-made information density on Earth is produced by the high energy accelerators at CERN. But these accelerators have to be $\sim 10^{15}$ times more powerful to reach the possible ultimate unit of information corresponding to the Planck length of a staggeringly small 10^{-35} meters.

Verlinde heroically *postulates* that the information of dark energy at the scale of the Planck length fills the whole Universe and is entangled, making the universe one big information processing machine. The jury is still out whether his theory is correct, but at least his predictions for dark matter distributions are consistent with our lensing results with OmegaCAM[28]. A theory of this kind seems promising, as it seems to solve the three most outstanding problems in physics:

1) the unknown connection between the theories of Einstein's general relativity and quantum mechanics,
2) no detection so far of the dark matter particle and
3) no understanding of the nature of dark energy, which accelerates the expansion of our Universe.

Unfortunately, there is no way to reach this unit of information with our particle accelerators. This enormous gap in reaching all the domains in the Information Universe is sobering — an instructive table is in the Appendix.

However, nature provides us with incredible laboratories in deep space, allowing us to observe extremely high energy events, such as the bending of light around black holes and the gravitational waves caused by merging black holes and neutron stars.

Here, we observe domains of the Powers of Two beyond 2^{128}.

CHAPTER 4
DEEP SPACE

BY EDWIN A. VALENTIJN
and contributions by:
Manus Visser
Chris van den Broeck
Margot Brouwer
Erik Verlinde

E. A. Valentijn, *Powers of Two*,
https://doi.org/10.1007/978-3-030-58345-3_4

66 bits
$2^{66} = 7\times10^{19}$ states
9 exabytes

The Desert
66-400 BITS

UNREACHABLE

Most of the deep high energy universe is unreachable by human-built experiments as it exists in the domain beyond 128-bits — which in current **cryptography** is considered impossible to decipher without the key.

CERN LHC
Credit CERN LHC

At CERN (Geneva) scientists are searching for highly energetic fundamental particles and ultimately for a Grand **Unified Theory** (GUT) which is extremely remote in **information density** space.

CERN's Large Hydron Collider (LHC) in Geneva is currently the record holder in acquiring huge amounts of experimental science data. The reason is simple: the target particles are extremely rare and are searched for in an ocean of other events. In 2019 CERN's advanced tape storage system Castor held 330 petabytes, in a very sustainable way compared to the energy-consuming spinning disks. CERN is already planning for tape systems two times bigger in 2022, and five times bigger in 2026.

Next to the tapes, the disk-based distributed storage system of 200 petabytes was copied around four times to other sites over the world to deliver 830 petabytes. Assuming that future data will be copied four times to other sites, CERN is actively planning for and implementing 9,000 petabytes, which is an amazing 9 exabytes.

This is equivalent to the number of **states** of a 66-bit string. In Information Universe speak, one could say that CERN's LHC experiment works at the 66-bit level. This is amazing but also sobering, as the ultimate limit in physics is at the Planck length, which requires around a 400-bit string to address all states in the universe. This enormous domain from CERN's 66-bit to 400-bit is called '**the desert**'.

The number of states scales with a typical size in meters of the information-carrying unit. This is depicted on the right side of 't Hooft's graph. On the left side the scale indicates the energy expressed in tera-electronvolt (TeV). The desert stretches up to 10^{15} times the current maximum energy of particles observable at CERN. Unreachable. The hope is that these high energy particles leave some trace at lower energies, detectable by accelerators.

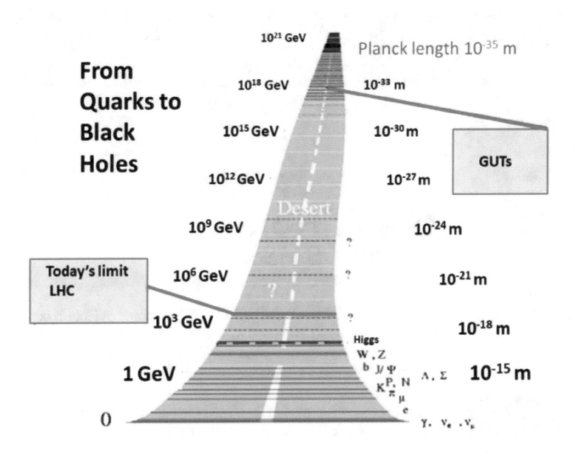

From Quarks to Black Holes

10^{21} GeV

Planck length 10^{-35} m

10^{18} GeV

10^{-33} m

10^{15} GeV

10^{-30} m

GUTs

10^{12} GeV

10^{-27} m

Desert

10^{9} GeV

10^{-24} m

Today's limit LHC

10^{6} GeV

?

?

10^{-21} m

?

10^{3} GeV

?

10^{-18} m

Higgs
W , Z
b J/Ψ
K P, N
π
μ
e

Λ, Σ 10^{-15} m

1 GeV

0

γ, ν_e , ν_μ

THE DESERT

"The ride through the desert" after 't Hooft 2003(!), showing the relation between the energy of a particle, from 1 up to 10^{23} giga-electronvolts (GeV), and its typical size in meters. In 2017 LHC's world record is 13 tera-electron-volts — even with ten times higher energies, there remains an enormous domain of unexplored information space: the desert. If a dark matter particle exists, it might hide there.
Credit Gerard 't Hooft

Detecting the Higgs Boson in the LHC in an ocean of other collision remnants at the CERN-Atlas experiment.
Credit CERN

Information density
The Desert
Cryptography
States
Dark matter
Unified theory

1001 1000 1100 1100 0011 0110 0011 1011 0110 0011 1100 0011 0110 0011 1011 0110
0011 1100 0011 1100 1100 1101 1011 1110 1001 1000 1100 0011 0110 0011 1011 1100
0011 0110 0011 1100 0011 1100 1100 1101 1011 1110 1100 0001 1110 0111 1000 0110
0111 1000 1011 1000 1101 0001 1011 1110 1011 1100 0001 0101 1100 0111 0011 1000 01

MINDBOGGLING ENTROPY

How much information is stored in black holes? Do black holes destroy or leak information? A black hole whose mass is equivalent to our Sun's has an event horizon radius of 3 kilometers. Its entropy is mindboggling: 4×10^{77} or 2^{258} bits, a number equal to the total amount of particles in our Universe!

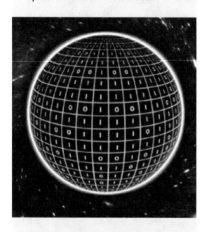

Remarkably, the mass of the black hole is fully determined by the amount of bits that can be projected on its surface. A tiny fraction of this information leaks out through Hawking radiation. A solar mass black hole has a temperature of only 60 nanokelvins, which means it absorbs more photons from the **Cosmic Microwave Background (CMB)** than it emits.

258 bits
$$2^{258} = 4 \times 10^{77} \text{ states}$$
in a 1 solar mass black hole

Black Hole
BY MANUS VISSER

Black holes are the most compact objects in our universe. They have extremely high **information densities** and actually store the ultimate amount of information space can host. If you try to store more information inside a black hole, by throwing in an object, its size will increase.

So how much information do black holes carry? Imagine a black hole as a spherical object. At a certain radius r, the gravity of the black hole is so strong that even light cannot escape: the event horizon. The surface area of this spherical boundary (given by: $4\pi r^2$) tells you everything there is to know about the black hole: its total information content or entropy S. This was first proposed by Jacob Bekenstein[29]. The entropy of black holes is simply:

$$S = \frac{k_B \, c^3}{G \, \hbar} \frac{\text{Area}}{4}$$

This magic formula is the "$E=mc^2$" of the black hole, and connects the information content or **entropy** in physical (in vivo) and communication (in vitro) systems. Strangely, this formula might also form a basis for a **unified theory**, as it contains the four fundamental constants of nature dictating the four very different realms of physics: thermodynamics (Boltzmann's constant k_B), relativity (speed of light c), quantum mechanics (Planck's constant $\hbar$) and gravity (Newton's constant G).

It is amazing that all the properties of a black hole relate to its surface area, rather than its volume. This is identified as the **holographic** principle: the information inside a 3D black hole volume can be described in one dimension less, on the 2D surface of the event horizon. The surface area is then measured in the smallest length unit we know of in physics: the Planck length = $1.6 \, 10^{-33}$ cm, the smallest unit of information in our Universe. Verlinde uses this principle to derive his **Emergent Gravity** and even **dark matter**.

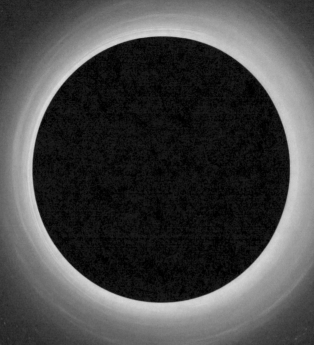

BLACK-BODY RADIATION

It is very satisfying that the laws describing black holes mimic those of particles in gas form. However, the most convincing evidence that black holes are thermodynamic objects is a famous calculation by Stephen Hawking. He showed that, if quantum effects are taken into account, black holes emit **black-body radiation**. Although classically nothing can escape from a black hole, quantum mechanically particles can tunnel through the event horizon. Black holes are not black after all. In fact, they slowly evaporate! For every black hole, this Hawking radiation is perfectly thermal: it has a standard black-body spectrum that does not contain any information about the specific black hole that emitted it, except its total mass. Therefore, it seems that the written information in this book would be lost when it is thrown into a black hole, since it cannot be recovered by analyzing the Hawking radiation. However, according to quantum theory, information is a conserved quantity that can never truly be lost. Physicists are still debating this "black hole information paradox", arguing whether

information can truly be destroyed by black holes. But what type of information is associated with black holes? It is fair to say that we do not yet know what the correct quantum description of black holes is. In string theory one can compute the entropy of certain idealized black holes from counting microscopic states. But for more realistic black holes, it remains an open question, from what type of information they are made.
Credit Warren Johnston

Black hole
Information density
4π
Entropy
Unified theory
Holography
Emergent Gravity
Dark matter
CMB
Black-body radiation

RINGDOWN

A newly formed black hole undergoing "ring-down". Spacetime in its vicinity is vibrating in a specific way. In the next few years, as detector sensitivities improve, it will be possible to study gravitational wave signals from ringdown in sufficient detail for us to tell the difference between the black holes of classical general relativity and modifications thereof, for instance as a result of macroscopic quantum effects.

Credit NASA/Goddard/UMBC/Bernard J.Kelly, NASA/Ames/Chris Henze, CSC Government Solutions LLC/Tim Sandstrom

UNRAVELING BLACK HOLES WITH GRAVITATIONAL WAVES

BY CHRIS VAN DEN BROECK

Detections of gravitational waves by LIGO and Virgo have become commonplace. In one case, the merger of two neutron stars was seen nearly simultaneously in gravitational waves and gamma rays, among other things showing that the speed of light equals the speed of gravity to one part in 10^{15}. However, most of the signals being detected appear to be from mergers of two black holes. These are perfect laboratories for testing the fundamental physics of gravity: here we observe empty spacetime whose curvature is a million times stronger than in any other astrophysical scenario, and is evolving on a timescale of milliseconds. This has given us access to a regime of gravity where Einstein's general relativity does not just provide minor corrections to Newton's theory of gravity; instead it takes center stage. For now, Einstein's theory holds up.

The next step is to answer the question: are the massive compact objects we observe really the "standard" black holes of classical general relativity? A range of possible alternative objects have been put forward, such as dark matter stars, or gravastars that resemble pockets of dark energy. Prompted by Hawking's information paradox, alternatives to black holes such as firewalls and fuzzballs were proposed. The latter are particularly exciting, since they open up the possibility of directly seeing quantum gravity at work.

When two black hole-like objects merge, a single, highly excited object is formed which undergoes "ringdown". This is similar to a chunk of metal being struck by a hammer: the object will vibrate with particular characteristic frequencies and damping times that are unique to its shape and composition. Observing gravitational waves from ringdown will enable us to tell the difference between Einstein's "ordinary" black holes and alternatives. For objects like fuzzballs, which don't have a horizon, there is also the possibility that gravitational waves falling in will bounce around many times inside the object, at regular times leaking out again as gravitational wave "echoes". Should such anomalies be observed, it would be a watershed moment in our fundamental understanding of gravity.

Gravitational waves
Black hole
General Relativity
Dark matter
Dark energy

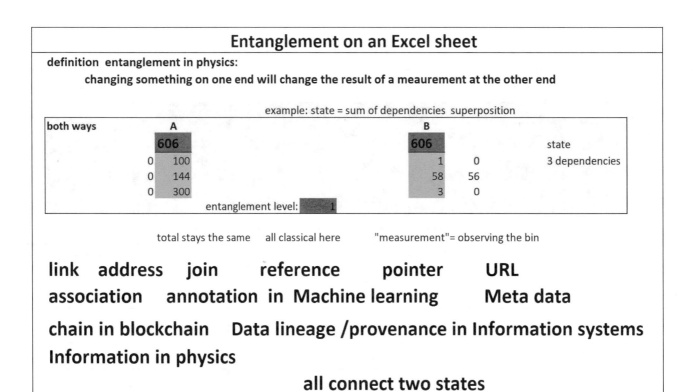

ENTANGLED STATE

With a simple equation one can create an entangled state on an Excel sheet (in vitro) — following the basic definitions/properties of entangled states in physics (in vivo), allthough the signal on the Excel sheet is classically transmitted by electrons in the computer hosting the Excel.

Snippet from my presentation at the Information Universe Conference 2018.

Credit Ron de Wit

Links — addresses
Artificial Intelligence
Cryptography
Data-centric
Open Science
Metadata
Astro-WISE
Virtual Observatory
DNA
Apoptosis
Entanglement
Complexity
FAIR
Standards
Unified theory

LINKS
THE UNIVERSE AS A SPREADSHEET

In Information Technology **links** are applied with an ever-increasing **complexity**. A link is nothing other than an address such as printed in old-fashioned telephone books. Addresses come in many flavours, but they all are fundamentally the same thing. Addresses are applied to programme computer chips (memory addresses), on the internet (URLs), in text and files (references), in databases (joins or associations) and in **Artificial Intelligence**, as annotations for data and to train neural networks. In blockchain technology, like the bitcoin, addresses are put in a block, which in turn are put into a chain, allowing computers to fully trace the sources. As we don't trust each other, everything is **encrypted** allowing only limited access to these links.

We use links to build whole worlds of complex information systems: in so-called **data-centric** data federations, **links** act as a workhorse to facilitate globally distributed storage and processing. For instance in the European Union, the Commission has recently decreed the **Open Science** paradigm: all research results and original data should become publicly available, following the **FAIR** principles — Findable, Accessible, Interoperable and Reusable. In fact, this implies that all these data should have administrated **links** and **Metadata**. Astronomers are quite advanced with such an approach (**Astro-WISE, Virtual Observatory**), but we know it takes dozens of years of work defining and implementing **standards** and protocols. It is a major, mainly sociologically oriented, project management task to achieve this for all fields.

There is no physical theory which describes what a **link** is: "we are just doing something" and thereby build huge systems (in vitro), which are like super spreadsheets. But in fact, nature (in vivo) does the same: e.g. a cell reads its **DNA**. By complex exchange of information with its environment — amongst others involving **apoptosis**, a programmed cell death — it determines and lives out its identity. In physics links are perceived as entanglement. This suggests we are hitting something very deep: **links** in Information Technology and even **entanglement** in physics appear to be a building block in the Information Universe — the in vivo – in vitro cross-over — is there any difference? In my view, a **Unified theory** should encompass information not only as a building block, but also as a carrier of links.

1001 1000 1100 1100 0011 0110 0011 1011 0110 0011 1100 0011 0110 0011 1011 0110
0011 1100 0011 1100 1100 1101 1011 1110 1001 1000 1100 0011 0110 0011 1011 1100
0011 0110 0011 1100 0011 1100 1100 1101 1011 1110 1100 0001 1110 0111 1000 0110
0111 1000 1011 1000 1101 0001 1011 1110 1011 1100 0001 0101 1100 0111 0011 1000

256 bits
$2^{256} = 10^{77}$ states
All matter

The CMB

BY MARGOT BROUWER

How much information does the entire Universe
contain? In every cube of 1.3 x 1.3 x 1.3 millimetres there is one
photon from the **Cosmic Microwave Background (CMB)** flying
around: here on Earth, in the space between planets, in between
the stars and galaxies — everywhere. It is amazing: CMB photons
are filling space everywhere in the universe. You can watch them
on the screen of an old analogue TV set with an antenna on
your roof — *see* page 18 Multicellular Life — although they were
emitted 13.8 billion years ago.

Where do all these **CMB** photons come from? At the dawn of
time, when our Universe was more than 1,000 times smaller than
it is now, all matter was compressed into a hot dense plasma
emitting **black-body radiation**. All photons continuously bounced
off the charged electrons like balls in a pinball machine, unable
to escape. Finally, 370,000 years after the **Big Bang**, the cosmos
had expanded and cooled enough for the protons and electrons
to combine into neutral hydrogen atoms. The photons were freed,
and able to start their nearly eternal journey through space. Now,
for each normal particle, there are billions of CMB photons. This
means that CMB photons totally dominate the information content
of our Universe, with a staggering 5×10^{77} units.

Still, the average separation of around 1.3 millimetres implies
that the average total **information density** of the Universe is
approximately equal to the text you are reading right now. The
overall (Information) Universe is rather empty! Obviously, we are
not counting unknown components, such as **dark matter**. In
his theory of **Emergent Gravity**, Erik Verlinde postulates that
dark energy carries even more information than the CMB.

256 LINES

Almost all the information
in the Universe is con-
tained in photons emitted
when the Universe was
just a new-born. A barcode
of 256 lines would have a
unique code (address) for
all of them.

Credit Creative Commons Zero

Credit Olcay Ertem

Comparing a 370,000-year-
old Universe to the present
is like comparing a baby of
one day old to a 100-year-
old woman.

ESA's Planck satellite view of the CMB all over the sky **Credit** ESA

BABY PICTURE

CMB photons escaped from the primordial plasma when the Universe was a baby of only 370,000 years old. This is a picture of that early moment. Now, 13.8 billion years later, we use the CMB to figure out what the early Universe looked like, particularly with how much structure (S_8) and visible+dark matter (Ω_m) it started. We compare this baby picture to other measurements, such as weak gravitational lensing in our local present-day, Universe (the old woman), to test our current cosmological model: ΛCDM. Until recently the data matched very well. But as we acquired more Big Data of galaxies in the local Universe with our own Kilo-Degree Survey (KiDS) (see page 67: From terabytes to two numbers), an inconsistency between the Planck and KiDS/DES surveys emerged in the values of S_8 and Ω_m. This might imply the need for physics which transcends General Relativity, perhaps in a new unified gravity plus quantum theory. Astronomers are quite excited about it, and ESA's Euclid satellite will shed more light on this.

I can't wait!

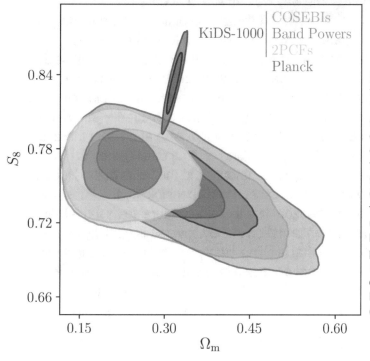

KiDS-1000 (2020) — the latest results from 1000 square degrees of Kilo-Degree Survey data, showing a discrepancy between their lensing measurements and the Planck CMB data, indicating that our present day universe has ~10% less structure than predicted by ΛCDM. **Credit** Marika Asgari et al. 2020 and the KiDS collaboration (arXiv:2007.15633)

Big Bang
Black-body radiation
Information density
Emergent Gravity
Dark matter
Dark energy
Unified theory
CMB
Lensing
Big Data
ΛCDM
General relativity

0011 0101 1001 1000 1100 1100 0011 0110 0011 1011 0110 0011 1100 0011 0100 0011 1011 0110 0011 1100
0011 1100 1100 1101 1011 1110 1001 1000 1100 0011 0110 0011 1011 1100 0011 0110 0011 0110 0101 1100
0101 1000 1100 1101 1010 1110 1100 0011 0110 0111 1011 0110 0011 1100 0011 1100 1100 1101 1011 1110
1001 1000 1100 0011 0110 0011 0011 1101 1001 0001 0011 0110 0011 1001 0111 1101 1100 1101 1001 1110
1100 0011 0010 1011 1001 0110 0011 1100 0011 1100 1100 1101 1011 1110 1001 1000 1100 0011 0110 010

2^{399}

In my information theory of **Emergent Gravity**, the information content of the entire Universe is 2^{399}: the highest power described in physics. This information is completely dominated by dark energy, and might help us to devise a unified description of the Universe.

Emergent Gravity from Quantum Information

"It from Qubit"

Gravity derived from 1st law of entanglement entropy

GR requires Area law => holds in Anti-de Sitter space. What about de Sitter space?

Erik Verlinde at the 2018 Information Universe conference.

399 bits
$2^{399} = 10^{120}$ states
Everything

Cosmic event horizon
BY ERIK VERLINDE

How much information is contained in our Universe? In 1998 astronomers discovered that the Universe is expanding at an accelerating rate. This means that, for galaxies beyond a particular distance, the space between us and these galaxies expands faster than the speed of light. Light emitted by these galaxies will never reach the Earth. So, similar to **black holes**, our Universe has a "Event Horizon" beyond which current events can never be known to us. To compute the amount of information inside the cosmic event horizon we can use the same **holographic** laws that Bekenstein and Hawking[29,30] discovered for the **entropy** of black hole horizons: the amount of information in the Universe is given by the area of the cosmic horizon measured in terms of the Planck length. This enormous number is of the order of $10^{120} = 2^{399}$ — a lot of information! But where is this information stored? Is there some way we can see or access it?

To explain the accelerated expansion of the Universe, one needs to postulate that space contains a kind of mysterious energy, called **dark energy**. One can think of dark energy as the energy of the vacuum: even empty space is filled with energy. In fact, in our current Universe dark energy accounts for close to 70% of the total energy. Dark energy can't be seen directly, nor can it be used in any way by human beings. Also at present there is no good theoretical understanding of what dark energy is made of. Yet, it gives the largest contribution to the value of the Hubble constant. Moreover, its presence is responsible for the appearance of a cosmological horizon. It is therefore natural to assume that the total amount of information in the Universe is actually carried by the dark energy that is filling empty space. Unfortunately this means that we can not access this information: it is hidden from us, just as the dark energy itself is. We can, however, use it in an attempt to build a **unified theory** of physics based on information.

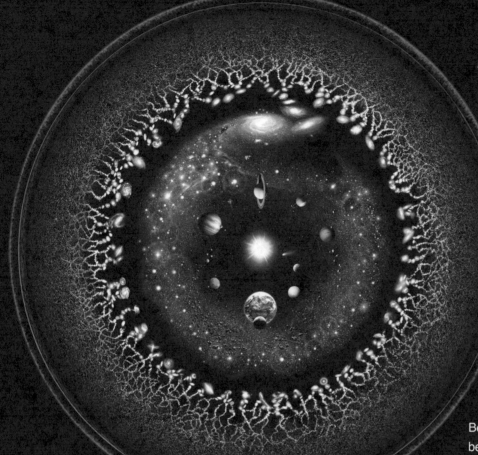

$$S = \frac{k_\mathrm{B}\, c^3}{G\, \hbar} \frac{\mathrm{Area}}{4}$$

Because the cosmological horizon behaves similarly to the horizon of a black hole, the total entropy (information content) of the Universe is given by the same equation: its area in terms of the Planck length. *Credit* Pablo Carlos Budassi

NEW THEORETICAL UNDERSTANDING

The presence of dark energy adds an enormous amount of information to our Universe, which has to be taken into account in deriving the gravitational laws. The additional information associated to dark energy leads to modifications compared the gravitational laws derived by Newton and Einstein at galactic and cosmic scales. In fact, one can show that these modifications have precisely the correct size to explain the phenomena observed in galaxies (and clusters) that are currently attributed to **dark matter**.

The large amount of information associated with dark energy also influences the structure and evolution of the Universe at cosmological distances and very early times. It thus appears that a new theoretical understanding of our Universe is emerging, which is based information and promises to shed a completely new light on the origin of spacetime, matter and all forces of Nature.

Black hole
Holography
Entropy
Dark energy
Unified theory
Emergent Gravity

Credit F. Pastawski, B. Yoshida, D. Harlow en J. Preskill.

Credit Angela Miller

MISSING MASS

Space itself can be described as a network of tiny "bits" of information, all entangled with each other. This quantum **entanglement** of space is what allows us to move from one place to another. The amount of shared information (entangle-ment **entropy**) between two volumes is defined by the surface area that separates them. As we just saw, this *area law* ($A \sim R^2$) is essential to deriving Newton's and Einstein's laws of gravity ($F \sim R^{-2}$). In my view, it is not just space that has entanglement entropy — **dark energy** produces it too. Since dark energy is spread throughout space, this entropy is not related to its surface, but to its *volume*. Around a mass distribution — such as a galaxy — part of this entropy is removed. The resulting "vacuum" pulls the surrounding information inwards, causing an additional gravitational pull. This extra gravity, caused by the interaction between normal matter and dark energy, exactly describes the behaviour of the "missing mass" currently attributed to **dark matter**!

It is observed in galaxies that dark matter only starts playing a role beyond a specific scale. In theories like Modified Newtonian Dynamics (MOND), this scale is seen as a new fundamental constant of nature. In my theory, however, this is simply the scale at which the information of the dark energy (described by a volume law: R^3) begins to dominate the information removed by matter (described by an area law: R^2). In my theory, this is no more funda-mental than the fact that a mouse can ventilate his body-heat more easily than an elephant. It simply shows the balance between area and volume, which causes stress at a specific scale. So, for the same reason an elephant has a sensitive and wrinkled skin (its volume/area ratio by-pass a critical scale), "dark matter" arises at large scales in the Universe. In this way gravity, quantum mecha-nics, dark energy and dark matter can all be described in one theory, based on perceiving the Universe in terms of information.

EMERGENT GRAVITY

BY ERIK VERLINDE

The biggest mystery in theoretical physics is: "How can we combine gravity (described by **general relativity**) and quantum mechanics into one **unified theory**?" Bekenstein and Hawking[29,30] found a profound clue in this cosmic mystery when they discovered that black holes can be described in terms of **entropy**, a concept derived from thermodynamics. So, just like the heat in a cup of coffee emerges from the combined movements of its molecules, gravity might emerge from the combined behaviour of the "atoms" of spacetime. If gravity, like heat, does not exist on quantum scales, the need for a unified theory dissolves! Now, how does **gravity emerge** from information? From **black holes** we know that the information in a spherical volume with radius R is equal to the number of **bits** N that fit on its surface area $A = 4 \pi R^2$. Each bit covers an area of a squared Planck length l_p^2 which can be expressed in terms of the fundamental constants as $c^2 / \hbar G$, so the number of bits on the surface of a sphere is:

$$N = A / l_p^2 = (4 \pi R^2 c^3) / (\hbar G).$$

Here c is the speed of light, $\hbar$ the Planck constant, and G... well, normally G would be interpreted as Newton's gravitational constant, but now it simply defines how many bits fit on a surface! Say that our volume contains a mass M with an energy $E = Mc^2$. How is this energy described by the information on the surface? Since the surface/horizon has thermodynamic properties, the bits behave like particles

in a gas: their total energy is equal to the total number N of particles (bits) in the gas times its temperature T and the Boltzmann constant k_B:

$$E = N k_B T.$$

By combining this with our first equation, we get a strange expression for the mass:

$$M = E / c^2 =$$
$$\tfrac{1}{2} N k_B T / c^2 = R^2 / G \, (2 \pi c \, k_B T) / \hbar.$$

What does this all mean? That will become apparent when we place a particle of mass m at distance R from our original mass M. Quantum physicist Bill Unruh discovered that a particle will undergo an acceleration **a** exactly equal to the yellow part of the expression for the mass M, because it experiences a type of Hawking radiation with temperature T. Thus we can simply derive the acceleration at the surface:

$$\mathbf{a} = G M / R^2 \longrightarrow F = m \, \mathbf{a} = G M m / R^2.$$

We have derived Newton's law of gravity, purely from the information properties of our system! Gravity is not one of the four fundamental forces; it emerges from bits like heat emerges from molecules.

Is the Universe one big information processing machine? In other words: is information the fundamental building block of our Universe? Is there a deeper It? Often this is referred to in J.A. Wheeler's (1989) **It from bit**: "the observer participance gives rise to information and information gives rise to physics".

There are several views on the fundamental role of information in Nature. In the previous part on Deep Space, we discussed Verlinde's view on **Emergent Gravity** and apparent **dark matter**, both based on the holographic principle of the amount of bits that can be mapped on a surface, and how the bits of **dark energy** are entangled in spacetime.

These and several other views are based on Wheeler's It from bit, which in turn reflects **Shannon's** formulation of bits and entropy in information and communication systems, presented in this Part 5.

But there are also other approaches, such as "Science from **Fisher information**" presented by Roy Frieden, with I being the information content in the observed phenomenology and J the intrinsic information content in nature. The information principle $I - J$ = minimal has the following specialized meaning: any information generally flows *from* a source, *through* a generally *noise-prone* (so-called 'lossy') channel and *to* a receiver (usually the observer). The quantity $I - J$ now represents the loss in information from input value J to output value I. Since that loss is minimized, it means that, overall, *the emergent information at the receiver is maximized*.

Part 5 also presents other views, such as Gottardi's **Tesserats** and Padmanabhan's **CosmIn**. In the end, our System of Science and consciousness is doing the job in settling what we believe to be true.

And then, there is the overwhelming diversity of the information content of **life**, trancending the typical astronomical numbers of the Powers of Two.

CHAPTER 5
IT FROM BIT

BY EDWIN A. VALENTIJN
and contributions by:
B. Roy Frieden
Stefano Gottardi
Thanu Padmanabhan

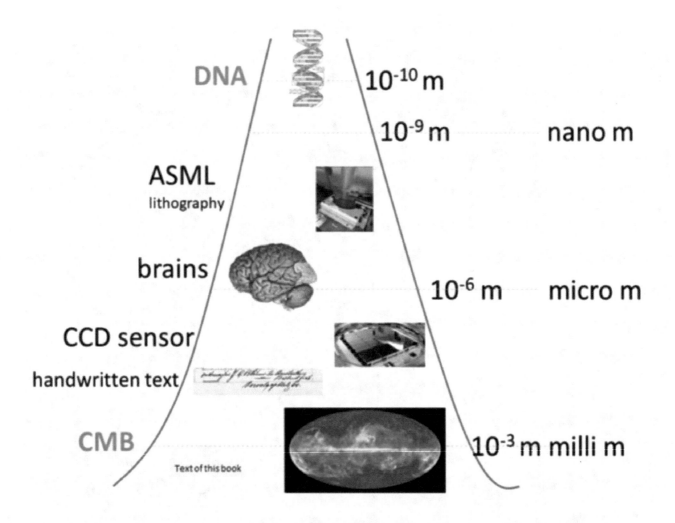

DNA 10^{-10} m

10^{-9} m   nano m

ASML
lithography

brains

10^{-6} m   micro m

CCD sensor

handwritten text

CMB 10^{-3} m milli m

Text of this book

SIZE OF INFORMATION UNITS

Typical size of units of information in our Universe, with large sizes mostly corresponding to low local densities at the bottom and small sizes / high densities at the top.

Remarkably, the highest *overall* information density in the cosmos is set by the Cosmic Microwave Background (CMB) and not by stars or black holes, but it still has a relatively low local density, comparable to that of the printed text in this book.

The figure illustrates our expanding universe with an ever-increasing entropy (disorder) and at the same time increasing organized structure (order) leading to increasingly complex systems and DNA micro biology.

The figure is continued on page 109 "The Desert" for the extremely high information densities.

INFORMATION DENSITY

How much Information is contained in space? How densely are bits packed? It helps to first to get an idea of the various **Information densities** in our universe. An easy way to do this is to consider simply the average size of a certain bit.

For instance, the pixels of the **OmegaCAM** CCD sensors are 15 micrometers (1.5×10^{-5} meters) in size. At this dimension information is collected. In 2019, the computer chip manufacturer ASML started to use extreme ultra-violet light (nearly X-rays) to print the patterns in the silicon wavers of computer chips. The light has a wavelength of 13.5 nanometers, reaching an information density on state of the art computer chips of one per 1.3×10^{-8} meters.

Our brains contain around 89 billion nerve cells (neurons), connected by trillions of synapses[31]. When single neurons contain basic units of information this corresponds to a density of one unit per 2×10^{-5} meters. But this is a minimum estimate, as the trillions of synapses could contain much more information. Hence, the information density in the **human brain** is estimated in the range of a bit per $2 \times 10^{-5} - 2 \times 10^{-6}$ meters.

The human **DNA** contains 6 billion bases on a string which amounts to around 1 meter when the tangled string is unfolded. With 1.7×10^{-10} bits per meter, DNA is a powerful topper as it works on a molecular level.

The average information density in our universe is totally dominated by the radiation of the **Cosmic Microwave background (CMB)**. It is much higher than the average information density from particles — simply because the universe is very, very empty. Amazingly, the CMB is the only component of the universe which is everywhere — it fills space: in roughly every millimeter of the Universe there is one CMB photon — but still it is at the low end of the density diagram on the left page.

Remarkably, the overall figure follows the fundamental trend of our expanding universe, with an ever increasing entropy (information spread over the largest amount of space) in the CMB, while at the same time highly organized structures are formed with a high information density, topped by human DNA.

Information density
OmegaCAM
Human brain
CMB
The Desert
DNA

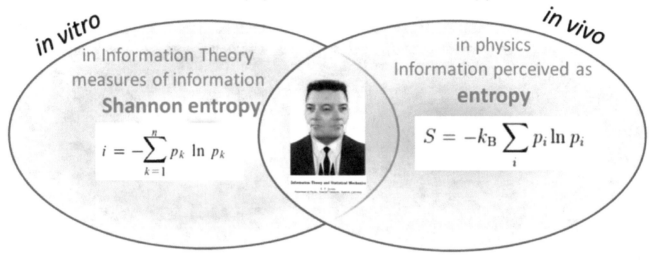

Information both in physics and ICT modeled as entropy

in vitro

in Information Theory
measures of information
Shannon entropy

$$i = -\sum_{k=1}^{n} p_k \ln p_k$$

in vivo

in physics
Information perceived as
entropy

$$S = -k_B \sum_i p_i \ln p_i$$

Information Theory and Statistical Mechanics

E. T. JAYNES
Department of Physics, Stanford University, Stanford, California

E.T. JAYNES

Both in physics (in vivo) and in Information Theory (in vitro) the total amount of information in a confined system is perceived as entropy: S in physics, i in Information Technology.

In both cases the entropy is equal to the sum of the probabilities p of all states that the particles of that system can possibly have. The exception is the Boltzman constant **k** in physics, because here the system has a temperature.

E.T. Jaynes[32] (1957) *in the middle of the figure* was the first to discuss the remarkable similarity between both expressions, but noted that this does not prove they have the same cause. Still today, this is a key item in understanding the cross-over of the in vivo – in vitro Information Universe.

Another key notion was presented by Shannon[24]: Information only has meaning when it is being copied. In other words Information only exists in relation to its environment by being copied. Indeed IT (Information Technology) equals ICT (Information Communication Technology).

ENTROPY

What is information? In physics (**in vivo**) information is perceived as the **entropy** of a thermodynamic system, which has a temperature. Entropy is often referred to as the measure of disorder in a system, and it is determined as the sum of all the *possible* **states** of the particles in a system, such as a confined gas or a fluid in a glass. All these possible states can add up to huge numbers and it reflects the total disorder in a system.

Our Universe has the very fundamental property that the overall amount of disorder, entropy S, is continuously increasing, while at the same time the amount of organized structures such as stars or anything composed of particles is continuously increasing. This is truly one of the most fundamental and miraculous properties of our Universe: while highly organized structures are continuously growing, the total amount of disorder is also continuously increasing. This is so fundamental that we have no idea why this is happening, and we might revert to the **Anthropic principle**.

Our universe as a whole, its entropy, and thus the amount of information, is totally dominated by the photons of the **Cosmic Microwave Background (CMB)**, which is around 5×10^{77} bits, the largest truly observed number in our Universe.

Heroically, Verlinde postulates that **dark energy** has a temperature and hence contains information which leads to a staggering 10^{120} units, the largest number we know about — it requires a 399-bit pointer to identify all states.

In Information theory (in vitro), **Shannon** was the first to formulate entropy in information systems in his famous paper from 1948[24]. But first of all, he noted that information only exists when it is sent (copied) from a source to a receiver. Information Technology is communication. The entropy is obtained by summing the probabilities of all possible events. Think of a buoy on the ocean, with a sensor that measures the waves: highly probable events such as the daily effects of the tides do not contain much information, while unlikely events with low probabilities, such as a tsunami, carry a high amount of information.

Entropy
States
Being copied
Anthropic principle
CMB
In vivo
Dark energy
Shannon

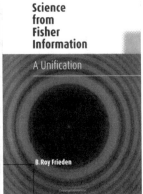

Science from Fisher Information

A Unification

B. Roy Frieden

This interchange between the observed universe *I* and the physical "an sich" universe *J* inspired Valentijn to develop Big Data information systems: "The Universe as a spread-sheet, page 83", "Astro-WISE extreme data lineage" and "Facts and Fakes", described elsewhere in this book. These systems are now also used to connect the cosmology of our Universe as observed by the Euclid satellite to the Big Data Euclid Archive System.

THE MAXIMUM FISHER INFORMATION

The fundamental idea behind my principle of maximum Fisher information is both profound and simple: all things in our Universe mutually interact in ways that are optimal. Information is transmitted through signals (e.g. light) that transmit maximum information (i.e. a minimum of information is lost), but with minimal change in the sender and the receiver such that the total system ultimately survives. Thus, the maximum Fisher information about a given signal is transmitted to a designated receiver, which similarly transmits the signal to the next, etc., to whatever final state exists in the given system. The ultimate resource (the 'medium of exchange') is the transmitted information. We show in a recent paper that, in biology, this gives rise to natural selection and evolution (*see also* page 44-45 by Lineweaver "where does biological information come from?"and the ear).

Where does this rule come from? Recall that all measurements are imperfect so that $y = \theta + x$. Therefore, there is some lower bound on the precision with which even multiple measurements can define a phenomenon. This is expressed by the Cramer-Rao inequality (developed in the 1940s to define constraints on statistical precision) which states the mean-square error *e* in the observational data from some phenomenon is limited by $e^2 \geq 1/I$, where *I* is the Fisher information. Thus, the level of Fisher information *I* in a system defines how well a given "state" value θ can be known

— the larger the *I*, the smaller the error. Requiring that *I* is always maximal gives rise to the basic functions that describe the system. This works for an extraordinarily wide range of phenomena: from quantum systems to growing biological cells, to the whole Universe and even to many universes. What does this say? Existence *is as it is* because it is an optimum — it gives maximum possible information about each independently *observed* phenomenon (system). And it is not that one system is optimized at the expense of the others, in the sense that a gain of information by one system is the result of information loss in another — they are all at optima simultaneously. What can we conclude from this?

1. Existence is an 'egalitarian' effect. No one system is a priori favored to the exclusion of others.
2. Everything lasts for a finite length of time (and space), i.e. all are "death" processes (at least, mathematically). Of course, being subjects of the effect, we emphasize its opposite manifestation: life and its opportunities, its potentials.
3. Many fundamentally *different* systems are governed by *the same* laws because the functional forms that describe each system all obey the maximum Fisher principle, and are thus subject to the same mathematical constraints. Also, maximizing the Fisher information is a smoothing effect, and many smooth curves are nearly the same (not so for abrupt, step-like curves).

SCIENCE FROM FISHER INFORMATION

BY B. ROY FRIEDEN

My books "Physics from Fisher information" (1998)[10] and "Science from Fisher Information" (2004)[11] define and develop a unifying principle of physics: that of 'extreme physical information'. The information in question is, perhaps surprisingly, not **Shannon** or Boltzmann **entropy** but, rather, Fisher information — a simple concept little known to physicists. Fisher information I connects the act of measurement: the phenomenon (Kant's "Erscheinung"), to the source in nature: the noumenon J (Kant's "Ding an sich"). It assumes that each measurement is subject to some random fluctuation, so that each measurement y of some phenomenon x is subject to a small degree of uncertainty θ. So, every observation is inaccurate such that $y = \theta + x$.

Now, J is the true information level in the phenomenon (e.g. of quantum mechanics), and the Fisher information I is the information actually obtained in an experimental observation. I then propose a general principle (the information principle) that in any exchange between nature and an observer, the information will be maximized. In other words, the loss of information in the exchange will be minimized:

$I - J =$ minimal.

In "Science from Fisher Information", statistical and physical properties of Fisher information are developed. This information is shown to be a physical measure of disorder, sharing with **entropy** the property of monotonic change with time. Information will tend to be lost over time in a thermodynamically closed system, but can increase in an open system. This information concept is applied to experimental observations to derive most known physics: from statistical mechanics and thermodynamics to quantum mechanics, the Einstein field equations, and even the law governing the growth of cancer mass m with time t: $m(t) = A \, t^\phi$ with $\phi = 1.618...$ (the Fibonacci golden mean).

Many new physical relations and concepts are developed, including new definitions of disorder, time and temperature. The information principle is based upon a new theory of measurement, one which incorporates the observer into the phenomenon that he/she observes. The 'request' for data creates the law that gives rise to the data — the observer creates his or her local reality. In fact, the observer and his/her **in vitro** technology connects and settles the **in vivo** universe — the many **crossovers** detailed in this book highlight the interchangeability between the two domains.

Two universes: ours (lower) and a neighbor (upper) connected by an 'umbilical pathway'. The pathway, which extends from somewhere within the upper universe to somewhere within the lower, is a Lorentz wormhole. The visualized portion is of unknown length but, for simplicity, is assumed to be infinitesimal, i.e. the two universes are side by side.
Credit Take 27 Ltd

MULTIVERSE FROM FISHER INFORMATION

BY B. ROY FRIEDEN

In a recent article[33] I — in collaboration with Robert Gatenby — show in detail how a **multiverse** of N universes forms from this informational principle: we propose that, initially, one universe exists or existed. How it comes to exist is currently unknown. Exterior to this universe is a nearby small reservoir of matter-energy. Overall, a physical scenario of *maximum total Fisher information* everywhere in existence is postulated. Therefore, when this universe comes in contact with some component of the exterior reservoir of matter and energy, the exchange of information will result in a second universe. Other universes can then be formed similarly — in a wave of creation — out of neighboring reservoirs of matter-energy. To ensure that the second and subsequent ones will successfully evolve cosmologically (and ultimately astronomically and biologically) they must have "built in" specific values of the 26 universal physical constants: these are *those known to exist in our current universe*. Hence, these values — information — are communicated from the assumed pre-existing universe to the second, such that the new one will successfully evolve cosmologically (and ultimately astronomically and biologically). The second, in turn, evolves as the first did by sending (copying) its 26 constants — now acting as gene-like "heritable information" — to a third, etc. In this way, a number of universes N is formed. This is a Guth **multiverse** of order N.

Each such universe experiences negative energy, and so initially expands outward during a smoothly accelerating process. In a simplest version this is purely an expansion in time on the usual grounds that, of all spacetime coordinates: **x, y, z** and **t**, only the *initial value* of time: **t = 0**, is well-defined as that of the Big Bang expansion era. Applying the principle of **maximum Fisher information** gives the resulting expansion: a probability law that is exponential with time — a known, standard result. In this manner a 'string' of N universes is formed, in any combination of parallel or serial existences. Note however, that in my opinion the above assumption of *simple expansion in time* is incorrect. This is on two grounds:

1. General relativity is four-dimensionally covariant, so we believe that instead the expansion is not only over time (as preceding), but over all three spatial dimensions as well. And so, just as time **t = 0** is a well-defined event, so is an initial position $x, y, z = x_0, y_0, z_0$. In principle, this initial position can be known. All such expansion processes are exponential in nature.

2. Also, the four expansions are statistically independent, which means their levels of Fisher information add. Thus, the total information in the 'multiverse' or order N is roughly 4N times that of the information in just the time-dimension of the single universe. This grand total obeys our premise of **maximum Fisher information** for the total system.

**Multiverse
Fisher Information**

Artistic representation of **tesserats** in the information reciprocal space (bottom) and galaxies in the real space (top). The two spaces are connected by a mathematical projection and they are equivalent representations of reality. This projection resembles a hologram.

Tesserats
General Relativity
Unified theory
DNA
Qubits
Non-locality

FUNDAMENTAL TESSERATS

In my view[34] Tesserats may turn out to be the fundamental building blocks of our Universe. This would have many implications for the interpretation and understanding of our current theories. For example, in quantum physics it is known that we may change the outcome of an experiment by simply watching. The theory of p-information describes interactions and ex- change of p-information among objects (watching) so that p-information is conserved. Once we are able to represent the true underlying physical reality of objects with p-information, properties such as their energy, momentum and mass should emerge without needing additional definition.

ON THE ORIGIN OF PHYSICAL INFORMATION

BY STEFANO GOTTARDI

The concept of information is profound and potentially more fundamental than the concepts of energy, mass and spacetime itself. We might argue that information is the key to a deeper understanding of the laws of Nature, opening the path towards the **unification** of physics theories like quantum mechanics and **general relativity (GR)**. But what is Information? Is it physical? If so, where is it stored? And how does Nature use it to build up the Universe? To address these questions, I started from key requirements that physical information should satisfy:

- should be quantized: it should be possible to count it
- should be finite: objects should be described by a finite amount of information
- should be conserved: it should not change if an object is "untouched"
- should be invariant: it should not depend on the observer.

From these requirements a definition for physical information is proposed: *physical information,* p-information, *is encoded by phases carried by physical waves and is quantized in four fundamental units: $1/2\pi$, π, $3/2\pi$, 2π.* Quanta of p-information may be named **tesserats** to distinguish them from the **bits** and **Qubits**. Surprisingly, tesserats are more similar to how **DNA** encodes information than to bits. In fact, they are equivalent to the complex numbers (i, -1, -i, 1) and thus carry orthogonality and interference properties. This means that they behave as if they were waves. According to classical physics, having quantized phases is just nonsense. However, in quantum physics we are used to surprises — we might have to quantize phases to make progress in our understanding of how the Universe works. More precise phases require more tesserats. As a consequence, more complex quantum waves (objects) require more p-information.

The Universe emerges as a projection from a reciprocal information space (where p-information "is stored") to the real space we live in. Information becomes **non-local**. Heisenberg uncertainty, wave-particle duality and quantum **non-locality** gain a straightforward interpretation in this theory.

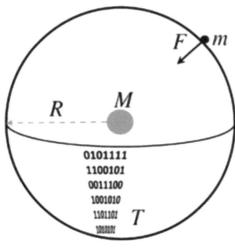

Verlinde's graph of Emergent Gravity *by mass M at the centre. According to Verlinde gravity is characterized by the maximum amount of information which can be transferred through the surface of a sphere M, at the centre of the graph, which dilutes with distance squared R^{-2}, like Newtons law of gravity.*
Credit *Erik Verlinde*

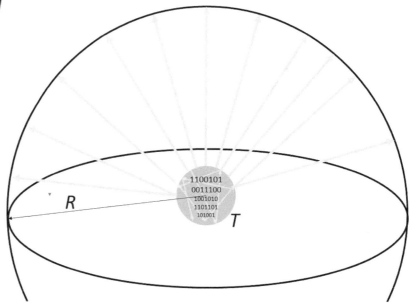

Black-body radiation emerging from the surface layer of the yellow sphere at the centre — within that surface radiation emerges in a digitized fashion. When leaving the surface it spreads maximally over space, and dilutes with the distance from the surface squared ~ R^{-2}.

SPREADING INFORMATION OVER SPACE

The amount of information spread over space by the central body is in both cases (black-body and EG) determined by the maximum amount of information that can be mapped on or "pumped" through the surface. For black bodies this amount is set by the temperature T of the surface and the discrete energy packages $E = h\nu$ of the photons. For entropic gravity (in the case of Verlinde's hypothesis) this is described by the holographic principle saying that the maximum information content of a surface, in bits, cannot be more than its area measured in Planck length, squared.

After leaving the surface the two seemingly unrelated physical processes of black-body radiation and gravity follow a so-called R^{-2} surface law, meaning that both dilute with distance from the source squared. Though a unified theory that combines General Relativity (gravity) and Quantum Mechanics (black-body) is not yet identified, the remarkable correspondence of both the information flow through surface and the dilution with distance seems to hint at a theory based on information. The riddle of unification is not solved yet, but information as the fundamental building block seems the best candidate for the glue. In any case, such a unification would have the consequence that our universe has, as its most fundamental and overall property, to spread information over a maximum amount of space — already from the very beginning.

BLACK-BODY RADIATION

Every object radiates and cools down when it is placed in a cooler environment: this is called **black-body radiation**. Earth cools down at night. In 1900 Max Planck was the first to design the formula describing this radiation of photons. His work is considered as one of the three big breakthroughs in physics in the last century, next to Einstein's General Relativity and quantum mechanics. His formula is very well verified, however it contains some mysterious aspects that Planck originally labelled as "an act of desperation".

He assumed that the energy E of photons comes in very discrete steps, (quantized) $E = hv$, with v the frequency of the photon. Though this principle is used everywhere in physics, the nature behind this expression for the energy of a photon is yet unknown. E.T. Jaynes[32] commented: *"E = hv is in present quantum theory only an empirical relation for which — astonishingly — nobody seems ever to have sought a theoretical explanation; we expect that a full understanding of E = hv will be of just as fundamental a nature as is General Relativity, far beyond anything in present quantum theory".*

Planck went on, combining two fundamental principles: Maxwell's finding that the wavelength of a photon is related to its frequency v as c/v with c the velocity of light; and that only standing waves of specific wavelength survive at the surface of bodies. From this combination he derived the formula for black-body radiation. The mysterious quantised energy of photons and the discrete wavelength of the standing waves can both be taken as the source of the radiation in units of information — information as the fundamental building block.

Maxwell's notion of expressing wavelength in meter per wave justifies our using of **information densities**: bits per meter. For example, the extreme UV photons, with a wavelength of 13.5 nanometer, used by ASML to etch wafers to produce computer chipsets represent an information density in vivo of the light beam, but simultaneous also in vitro on the computer chip; an ultimate **cross-over** in vivo – in vitro, also called in silico in this context. At this point I wonder whether there is any fundamental difference between the two domains.

Holography
Unified theory
Black-body radiation
Information density
Cross-over
Emergent Gravity

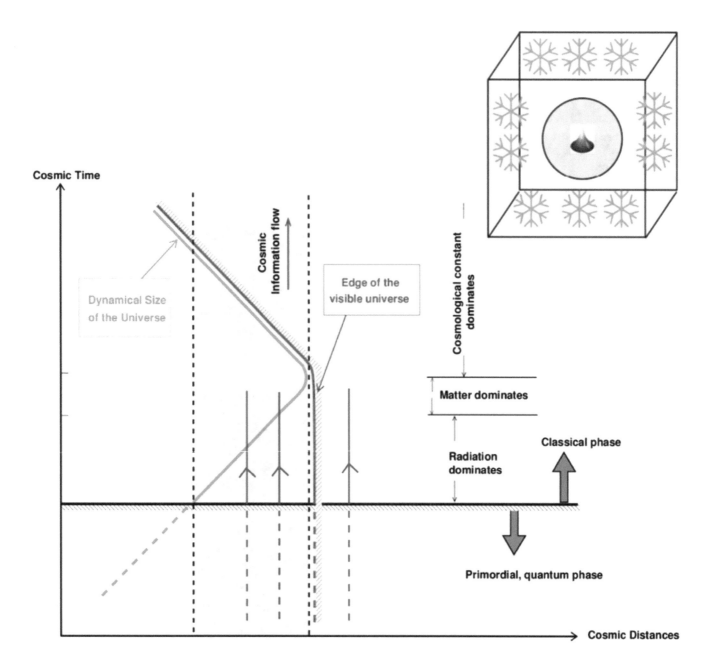

CosmIn

Flow of information in the Cosmos. The figure shows the different phases in the evolution of the universe. The horizontal black line divides the quantum pre-geometric phase of the universe ("ice") from the classical phase ("water") described by Einstein's theory. As the classical universe evolves it goes through radiation-dominated and matter-dominated phases and ends up in the current, accelerating phase, dominated by the cosmological constant Λ. The edge of the visible universe is shown by the red line and the dynamical size of the universe ("Hubble radius") is shown in green. The blue band denotes the length scales relevant for the flow of information from the pre-geometric phase to the classical phase. The amount of information — CosmIn — determines the value of the cosmological constant Λ.
Credit Thanu Padmanabhan

COSMIC INFORMATION COSMOGENESIS AND DARK ENERGY

BY THANU PADMANABHAN

What was before the **Big Bang**? The approach based on accessibility of Cosmic Information leads to a fascinating new picture of the universe: one that is very similar to a chunk of ice being heated by a point source at its centre. The Universe began with a **Big Melt**, like a heat source melting the ice of **Ti** (*see* page 10 "Ti") from the inside! The region close to the point source has melted and reached thermal equilibrium, while the region farther away, near the outer boundary, has not yet done so.

The observed universe obeys Einstein's equations and is surrounded by a pre-geometric phase that obeys the as-yet unknown laws of quantum gravity. The quantum-to-classical transition occurs at a very high energy thereby replacing the Big Bang. The two phases are connected by the amount of Cosmic Information — **CosmIn**. The concept of Cosmic Information is illustrated by this analogy: when a piece of ice melts to form water the total number of atoms in the ice will be the same as the total number of atoms in the water. Similarly, the phase transition which led to the birth of the Universe can be described by a number which relates to the degrees of freedom in the pre-geometric phase with those of the classical spacetime. Hence, introducing the CosmIn, a conserved quantity, which measures the total information transferred from the quantum gravitational phase to the classical phase of the universe, allows us to connect these two phases in a fascinating manner, bypassing the complications of a complete quantum gravity model.

Quantum gravitational considerations advocate an astonishingly simple value for CosmIn at the moment of the transition: **4π,** the number of information 'bits' on the surface of a sphere of radius equal to the Planck length. Using this and the CosmIn one can relate the numerical value of cosmological constant **Λ** — the simplest possible explanation for **dark energy**, possibly *the* deepest unsolved problem in theoretical physics today — to the energy scale at which the universe made the quantum-to-classical transition.

This is in turn related to the amplitude and shape of quantum fluctuations which acted as seeds of cosmic structure. This is the first time that a model with no adjustable parameters is able to provide a holistic explanation for both these observations, which has far-reaching implications for the quantum structure of spacetime.

Dark energy
CosmIn
Ti
Big Bang
Big Melt
4π

THE COSMOLOGICAL CONSTANT

The very existence of the cosmological constant, as well as its tiny value, can be understood as a direct consequence of the information content of cosmic spacetime. As a bonus, the analysis also leads to the correct value for the size and shape of the small fluctuations in the early universe which acted as seed for cosmic structure.

Credit Hamsa Padmanmabhan, edited by the graphic designer

CosmIn and the Cosmological constant

BY THANU PADMANABHAN

Observations tell us that dark energy, which is today speeding up the expansion of the universe, can be explained by introducing a term called the cosmological constant Λ (Lambda) into Einstein's equations of gravity. But for this explanation to work Λ must have a very specific and enormously tiny value: one divided by a number written as one followed by 123 zeros, which is in the domain of supposedly unbreakable cryptography. Explaining this value is considered one of the greatest challenges faced by theoretical physics today.

During the creation of the Universe, the pre-existing information in CosmIn determines the value of the cosmological constant Λ that accelerates the expansion of the universe at late stages. CosmIn, being a physical observable, must be finite. It will be finite only if the universe undergoes an accelerated phase of expansion at late times, exactly as we observe today. This is because a universe undergoing accelerated expansion at late times driven by a non-zero cosmological constant will possess what is known as a cosmic horizon. Such a horizon will ensure that even an eternal observer can only acquire finite amounts of information through signals from the past. One can write a precise mathematical relation connecting the amount of information an eternal observer can acquire and the numerical value of the cosmological constant. Thus, this connection not only suggests a fundamental reason for the existence of the cosmological constant, but also a means of calculating its numerical value, in terms of CosmIn. The value of CosmIn in the pre-geometric, quantum, phase can be determined using a result which repeatedly crops up in different models of quantum gravity: the spacetime behaves like a two-dimensional system when probed close to the Planck scale. This leads to the result that the total information transferred from the pre-geometric phase to the classical phase must be equal to a simple number: 4π, just the area of a sphere of unit radius.

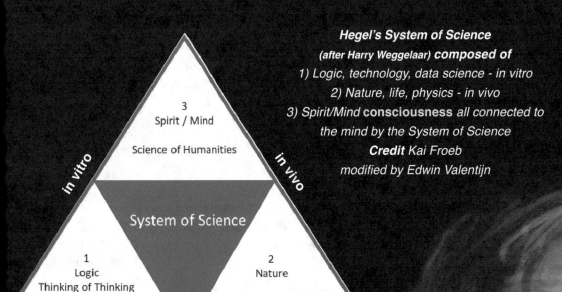

Hegel's System of Science
(after Harry Weggelaar) composed of
1) Logic, technology, data science - in vitro
2) Nature, life, physics - in vivo
3) Spirit/Mind consciousness all connected to
the mind by the System of Science
Credit Kai Froeb
modified by Edwin Valentijn

HEGEL'S ROLE FOR THE MIND

Hegel's "Logic and Nature" maps directly onto the in vitro — in vivo Information Universe as painted in this book. Hegel's role for the mind is rather unique: Man is slowly becoming God in creating his own world: the System of Science connecting Logic and Nature. Every step of newly acquired knowledge (3 Mind) brings mankind closer to being God. Hegel did push this to a megalomaniac extreme, particularly because he considered the worlds of Logic and Nature as alien. Indeed, quantum theory gives the human being a divine role when observing the universe and settling in which one she is living. Moreover, our understanding of consciousness moves more and more to the domains of Logic (in vitro) and Nature (in vivo) with a prominent role for links and associations. So after all, the Hegelian view of mind and consciousness eventually becoming submerged by Logic and Nature seems timely.

But how? I'll discuss this on the pages consciousness. *Painting by Mitchell Nolte*

In vitro
In vivo
Links — addresses
System of Science
Eyes
Consciousness
Facts and Fakes
Moore's law

SYSTEM
OF SCIENCE

What is the System of Science?
Scientists work with a list of basic rules.
Given these basic and new working rules,
eventually conclusions in research are
obtained. Often, students and junior
researchers get these rules in passing (so
not deliberately) from their tutors. Hegel
was one of the first to recognize this as
a closed system and described this in
his work "System der Wissenschaft, Die
Phanomenologie des Geistes" published
in 1807. Though more then 200 years
old, his work is actually very applicable
to the contemporary way of reasoning,
doing and reviewing science. It is a closed
system describing what and how, given
the rules, and there is no outside, no
way out, no why and origin. As Friedrich
von Schelling said: "it starts with light;
it does not create light out of darkness".
All objections to the system are part of
the system. It's the scientific system.
It does not explain origins, be they of
an electron, charge, magnetic field or
information. This notion should not be
taken as a ticket for a free ride in so-called
Metaphysics, but it helps to appreciate
the limitations of our scientific system;
we use our eyes and our consciousness
to describe the functioning of our eyes.
Our acceptance of reasoning with closed
systems legitimizes also alternative
closed systems, which happens in
different cultures, but even within the USA
— see Facts and Fakes, page 157.

Actually, the scientific system is
highly successful, particularly due
to another important feature of the
Information Universe: "Parents give
their knowledge to their children", or
better "generations to generations".
Evolution by knowledge transfer —
copying information — is so much
faster than by natural selection a la
Darwin. Also Hegel studied the classics
from Aristotle to Spinoza, Voltaire
and Rousseau to reach his views.
Information is copied from the Greeks
all the way to us. Of course, we all
know this, but it is such a dominant
process, with us in the middle, that we
tend to take it for granted. "Alternative
systems using alternative facts" are
often run by smart people without
education, i.e. without copying
information from previous generations.
To illustrate the incredible power of
knowledge transfer: economists have
no clue why the current inflation rate in
the economic world is so low; I bet it is
due to the hyper information copying,
which currently is taking place, and
makes things work so much better for
less money — a single smartphone has
more computing power than the Apollo
missions to the Moon had on board
50 years ago. While the annual increase
of salaries is of the order of a few percent,
the number of digital pixels on our
screens increase annually on average by
40%, following Moore's law: information
technology doubles its performance
every two years. Our knowledge evolution
is in many fields $2^3=8$ times faster than
the inflation rate — no surprise that this
power can break the inflation, salary
and economic growth circle, eventually
leading to negative interest rates. It's
a small price to pay for the enormous
information accumulation.

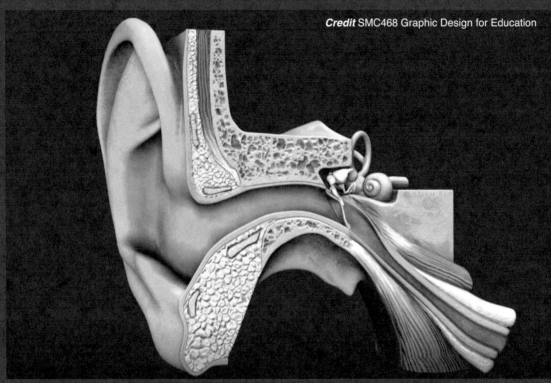

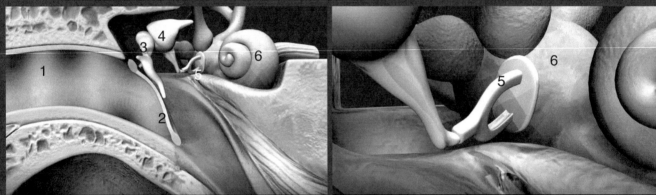

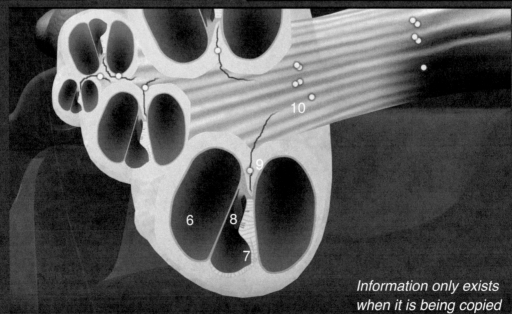

*Information only exists
when it is being copied*

THE EAR
COPYING INFORMATION

The path of the information flow of sound through the human ear is amazing and consists of numerous steps in which information is copied from one medium to another.

First, the sound waves (1) in the air are channelled from the outer ear (pinna) through **the ear** canal and strike the eardrum which will vibrate — creating a copy of the sound (2). In turn, this vibration is copied to and amplified by three tiny bones: the hammer (3), the anvil (4) and the stapes (5). The stapes copies the signal into fluid waves in the cochlea (6), and hair cells in the fluid react to the waves (7). Different sizes and chambers of the cells are sensitive to different frequencies, this way discriminating between high and low tones. Cells on top of the hairs copy the signal by producing ion-/chemicals (8) whose signal is translated by neuro-transmitters (9) into an electrical signal. This signal streams to the nerves towards the brain (10) where it's eventually copied and interpreted leading to an awareness by the **consciousness**.

This example demonstrates quite a number of characteristics of information. First of all, information seems only to exist when it is **being copied**. Although, the example of the ear is obviously **in vivo**, this also holds for ICT — Information and Communication Technology (**in vitro**) — a perfect **cross-over**.

Secondly, the copying process in the ear consists of many different physical processes, waves in the air (1), mechanical vibrations (2,3,4 and 5), fluid dynamics (6,7), chemistry (8), electrical signals (9) and eventually the **consciousness** of the **human brain**. The same piece of information seems to be carried by all these different and basic physical processes, suggesting a deeper and more overall role for information than all the specific parts of the flow, including waves, mechanical vibrations, hammers, fluid, thrilling hairs, chemicals and electrons.

Credit Still images from the video 'Anatomy of the Human Ear' by Stephen McHale.

The ear
Cross-over
Consciousness
Human brain
Being copied
In vivo
In vitro

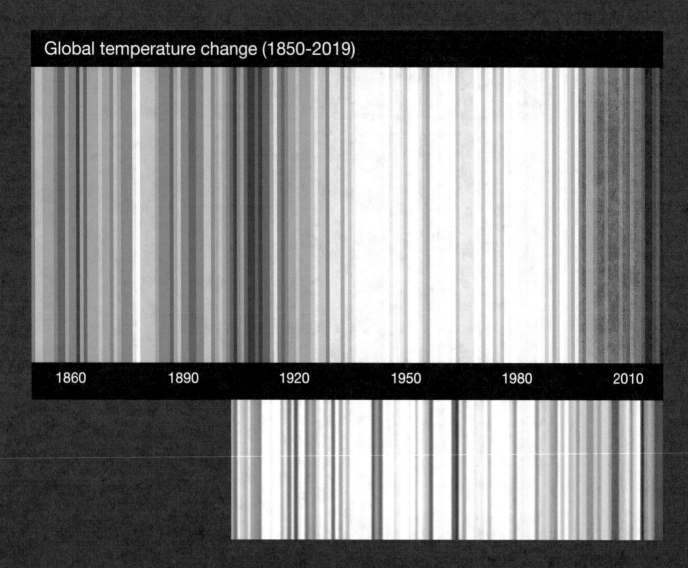

Global temperature change (1850-2019)

1860　　　1890　　　1920　　　1950　　　1980　　　2010

GLOBAL WARMING PICTURE

This iconic global warming picture from Ed Hawkins above the year bar shows in the blink of an eye the global warming on Earth. It depicts the average temperature per country, per year summed over 100s of countries on Earth. The colour scale goes from temperatures -0.7 °C (dark blue) to +0.7 °C (dark red), compared to the average during the period 1971–2000. This is a most powerful presentation of hundreds of millions of measurements collected over 100 years, and without fiddling with the colour scale it demonstrates the global warming. The message is made very clear without the need for any further details.
Credit Ed Hawkins (University of Reading) *see* https://showyourstripes.

Below the global graph: the average annual temperature in the centre of the Netherlands from 1901–2019 as recorded by the KNMI. The darkest blue corresponds to a temperature of 7.8 °C and the darkest red to 11.7 °C, visualising an increase of around 1.5 °C in the last 30 years.
Credit Karin van der Wiel (KNMI)

By comparing the graphs for different countries, adding a third dimension to time and temperature, one can directly observe that the warming up in the last 30 years is truly global.

VISUALIZATION
FROM IN VIVO TO CONSCIOUSNESS

Though **the ear** contains a marvelous system to dynamically copy information into the **human brain** and **consciousness**, most of the information gets to the brain by means of the **eyes**. The ear–mouth (aural–oral) axis allows more dynamic bi-directional communication; visual communication relies on facial expression and body language, which is more limited. I understand that deaf people feel much more isolated than blind people do. But the eye is marvelous at getting complex data to the brain. In this book I tried to relay a lot by means of visualizations in pictures — actually, one of the original drafts was made in the form of a movie story board. But perhaps a version with music and a moderator in a planetarium show would be optimal!

For the analysis of Big Data collections, visualization is absolutely critical. Mostly this is done on a 2D screen, but **CT scans** are 3D. At computer centers — such as that of my university — we operate virtual 3D environments, like a "cave" in which you can walk, or by using Oculus Rift devices. These environments are used for scientific discovery — mostly to search in datasets for unexpected events. As soon as we recognize these we quickly revert to 2D screens for further analyses. Also in planetariums, such as DOTLiveplanetarium which hosts the Information Universe conferences, we operate an immersive display with active stereo shutter glasses to view a semi-sphere in 3D.

But even 3D is not enough, as many datasets contain numerous parameters, and we often do not know *a priori* which ones are important and which ones are not. The visualization of multi-dimensional parameter space is still a major challenge in data science. This can be addressed with **Artificial Intelligence** type of approaches in which the machine determines the "feature vectors", i.e. the characteristics of a certain image — actually **Metadata**. But often this is done more successfully following human intuition — *see* e.g. Amina Helmi (page 56) on the discovery in 5D **Gaia** satellite data of stellar streams from an external galaxy merging into our Galaxy.

The ear
Human brain
Consciousness
CT scan
Artificial Intelligence
Metadata
Eyes

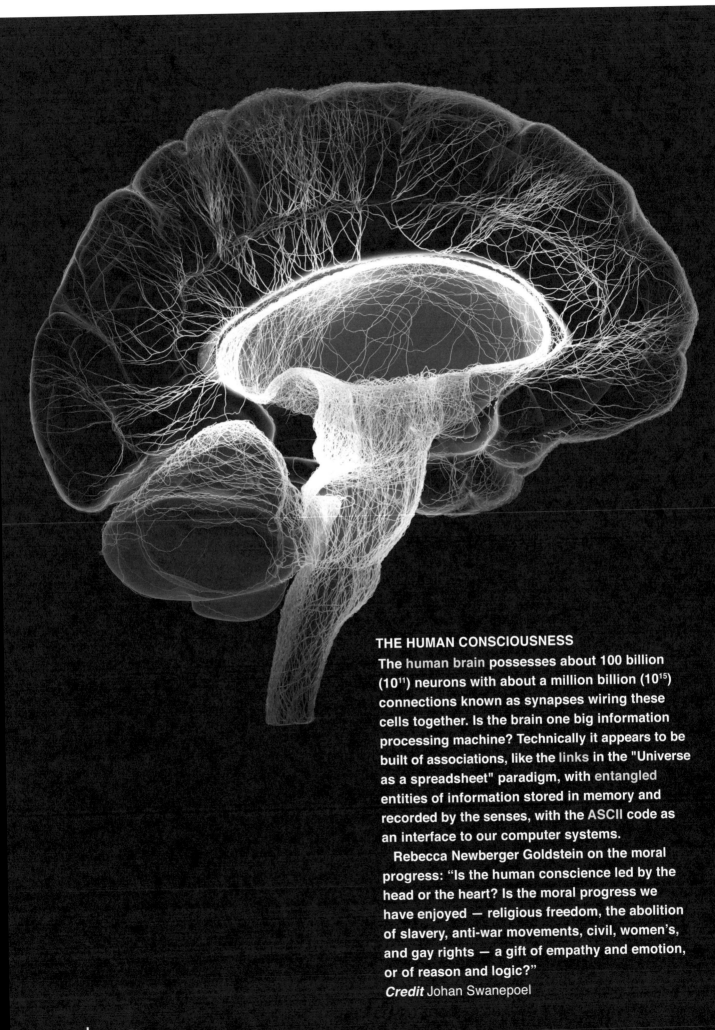

THE HUMAN CONSCIOUSNESS

The human brain possesses about 100 billion (10^{11}) neurons with about a million billion (10^{15}) connections known as synapses wiring these cells together. Is the brain one big information processing machine? Technically it appears to be built of associations, like the links in the "Universe as a spreadsheet" paradigm, with entangled entities of information stored in memory and recorded by the senses, with the ASCII code as an interface to our computer systems.

Rebecca Newberger Goldstein on the moral progress: "Is the human conscience led by the head or the heart? Is the moral progress we have enjoyed — religious freedom, the abolition of slavery, anti-war movements, civil, women's, and gay rights — a gift of empathy and emotion, or of reason and logic?"

Credit Johan Swanepoel

CONSCIOUSNESS

What is typical of **consciousness**? Some call the free will and the ability of "being alone with self" the most profound properties of the human consciousness. Very little is known about what this is, and not too much progress has been made. However, some well known "information" features of the consciousness need to be accounted for in any perception of the consciousness:

- in quantum mechanics things exist when they are observed by a human: the observer settles locally in which world she is living, which is actually an act of copying information from nature to the **human brain** and its consciousness (e.g. the paradox of Schrödinger's cat). Though in more modern views the human can be replaced.
- in quantum mechanics the observer affects the outcome of a measurement at a distance, *non-locally*, when states are **entangled** (e.g. **teleportation**)
- Scientists reveal **human brains** with extremely fast and complex networks of 10^{11} neurons, pumping around information and settling **links** between the information in memory and the outputs of the brains.

Brains seem to work like entangled information systems — processing, storing, linking and copying information. We call it in vivo, but technically it appears to work in vitro — the boundary gets very thin here, if it exists at all.

The question is whether our own will truly exists, or whether it is based on pure causality in the brain, based on associations (**links**) of past experiences such as events, language picked up from parents, influence from culture, etc. If one contemplates for a long time before taking a deep and important decision, at the very end, at the ultimate moment of the decision, it often feels like the decision just comes by itself, from an unknown source. It is plausible that the whole functioning of the brain and consciousness takes place in the domain of Logic and Nature — the ultimate cross-over of **in vivo – in vitro**, making this discrimination obsolete.

But, in **multicellular organisms** such as brains, the information exchange is mostly governed by mechanical and biochemical processes, not involving quantum mechanics. How can consciousness then play a role in the observation of quantum systems? We don't know — we could speculate that the **entanglement** of information in classical systems such as the brain is not different from that in quantum mechanics.

Human brain
Links — addresses
ASCII
Entanglement
Multicellular Life
In vivo
In vitro
Cross-over
Consciousness
Teleportation

INTELLIGENCE

On 21 December 2014 Ton Sijbrands repeated his world record in simultaneous blindfold draughts (checkers — US). In 48 hours he played against 32 opponents without seeing the boards, with 14 wins and 18 draws. Not all brains are equal, and he demonstrated the incredible power of the **human brain** when trained for a certain task. There has been a lot of attention to IBM's Deep Blue computer beating World Chess Champion Garry Kasparov in 1997, and Google's AlphaGo Master beating the world's top Go player Ke Jie, demonstrating the enormous progress of **in vitro** (in silico) **Artificial Intelligence** computing. But the notion can be reversed: the case of Sijbrands also demonstrates the enormous power of the network of the entangled information system in the human brain, built from associations / references — all different words for the same thing: **entanglement**. Actually, the case of Sijbrands is a simplified case of what we do on a daily basis. I guess Sijbrands does not even visualise the boards while playing — indeed like us in or daily life. The Hegelian view of mind and consciousness eventually being submerged by Logic and Nature. *Credit* Ton Sijbrands, ANP.

Human brain
In vitro
Artificial Intelligence
Entanglement

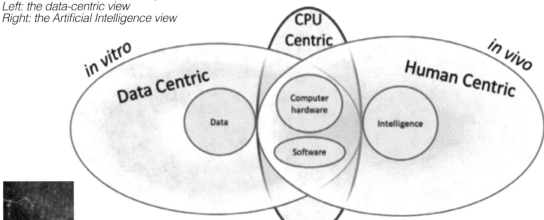

Three crosscuts of the Information Universe.
In the middle: the classical compute-centric view
Left: the data-centric view
Right: the Artificial Intelligence view

MACHINE LEARNING — CAUSALITY LOST IN TRANSLATION

The machine learning and the database oriented communities are often living on different planets.

People create, build and work with computers and write the software for the machines. While doing so they create abstractions in their mind, just to physically and mentally cope. Our abstractions started with **compute-centric** views, since originally the hardware of the processors were at the core of the thinking. As a student I had to bike every day to the computing center to deliver my job to the holy IBM360 machine. I organized the data (on punch cards) manually at home. In recent visits to large supercomputer centers, as at Jülich in Germany, I noted they still work according to this abstraction. A very computing intensive job is delivered to the machines, the job runs and the output is collected by the user.

But for data intensive computing this changed in the years after 2000 when digital sensors, such as CCDs, started to deliver data. Soon after that, data management became a much more demanding task than running the computers and a new era of **data-centric** views and abstractions emerged. In 2003 we celebrated the *first virtual light* (astronomers call the recording of a first image on a new telescope *first light*) for the **Astro-WISE** data-centric system, to be used for astronomical wide field imaging surveys. In 2008 I was asked to advise ESA on the merging of two satellite proposals into one, **Euclid**, delivering an unprecedented (for ESA) amount of sensor data, which has to be processed by eight data-centers spread over Europe and one in the US — the data-centric abstractions formed the core of the answer. I view the data-centric approach as the way to get to the source of the data, the raw data, nature, the Information Universe **in vivo** — actually enabling the **cross-over** in vitro – in vivo. It's such a natural approach, that a lot of extra benefits come along the way.

Interestingly, Tegmark[35] in his book *Life 3.0* discusses another abstraction, namely AI building both computers and software. This leads to a totally technically dominated abstraction, in which everything is run **in vitro** (in silico) and AI completely takes over the role of humans. It is rather science fiction. Karlheinz Meier, coordinator of the EU **human brain** project, emphasized at our second Information Universe conference that, in practice, creating anything like human intelligence in a computer would consume per CPU "transistor" 100,000 billion times more energy than the synapsis of neurons. Perhaps **quantum computers** can bridge the gap, but because they have to do a lot of computing to fact-check themselves, it remains to be seen whether this converges. Indeed, Life 3.0 seems a very long shot and far away from nature in vivo.

ARTIFICIAL INTELLIGENCE

While **multicellular life** can be seen as the first emergence of bio-intelligence at 2^2 and beyond, in vivo intelligence culminates in the **human brain** at 2^{36}. Computer code that mimics the network functioning of the human brain, **Artificial Intelligence (AI)**, was already developed in the 1950s. Since then the code improved, but much more important, the dramatic increase of both computer power and the amount of data storage following **Moore's law**, caused a take-off in the uses of Artificial Intelligence. When data with known properties are used to train an AI model, it is called supervised Machine Learning. In 1989, I wrote such a code to classify the morphology of the 31,000 digital images of the ESO-LV galaxies[7]. We had a training set with morphologies of 3,430 galaxies, determined by visual inspection by Corwin and De Vaucouleurs. I guess we were the first to do this, being the first to have such a large set of digital images.

Machine learning is good at doing tasks which take humans too much time, like the recognition of patterns in thousands of images. Already in our early programme we obtained accuracies of 80-90%.

Now a giant leap forward from 1989: 30 years later **AI** is hyping since the avalanche of Big Data requires machines to assist with the interpretation. It is very useful indeed, but there is a deep issue with machine learning: its working is a kind of black box. You never know how and why it came to a certain result. The only thing you know is that when the programme is trained and tested on known results it has a certain score, say: 90%, 99% or 99.9% of the times it is correct, with 99.9% being an exceptionally good score. For instance, the score of face verification by comparing for instance with a passport photo can vary from 90% in bad photo/light conditions up to 99.7% in ideal circumstances — Crumpler[39] .

This can be acceptable and useful for many applications, but when you want to discriminate **Facts and Fakes** this becomes a problem. The AI abstraction is opposed to the data-centric abstraction, the latter using databases, which **links** and **addresses** all the dependencies of a certain result to all the inputs — allowing the checking of the source, relations, etc (*see* p.83 "the Universe as a spreadsheet"). This checking cannot be done with results obtained with machine learning.

This is why I experience:
"Machine learning lost in translation", after Sofia Coppola's movie "Lost in translation" (Bill Murray 2003).

Multicellular life
Human brain
Artificial Intelligence
Compute-centric
Data-centric
Astro-WISE
Facts and Fakes
Links — addresses
Euclid
Cross-over
Quantum computers
Moore's law

Credit Bart Heemskerk

Jacco Gardner and Maria Pandiella performing at
DOTLiveplanetarium, Eurosonic music festival 2019

SOMNIUM LIVE

WITH JACCO GARDNER

Modern planetariums, such as the DOTLiveplanetarium in Groningen (NL), are fully digitalized. We display full dome with an **eJ** (not a VJ or DJ) managing the live show. The 360-degree imaging is obtained with eight projectors, a computer cluster connected to the internet and our large astronomy databases at the university computing center (*see* page 77: "Dark matter maps").

The virtual reality achieved **in vitro** in a dome seems to connect us to the **in vivo** world, often in an unforgettable way. When I conceived the Infoversum / DOTliveplanetarium I asked many people about their planetarium experience: often they could tell precisely what they saw more then 20 years ago. No way to properly illustrate this in a book or on a flat screen; what's in the dome stays in the dome.

Unforgettable is Jacco Gardner's edition of Kepler's Somnium in DOTLiveplanetarium, while playing his album with analogue synthesizers on our 64-speaker surround system.

Kepler wrote his novel Somnium in 1608, describing how daemons could **teleport** humans to the Moon and how they would observe Earth, planets and stars from there — actually following the Copernican principles, with the Earth rotating around the Sun. The narrative is referred to as the first science fiction book, long before Carl Sagan and Isaac Asimov.

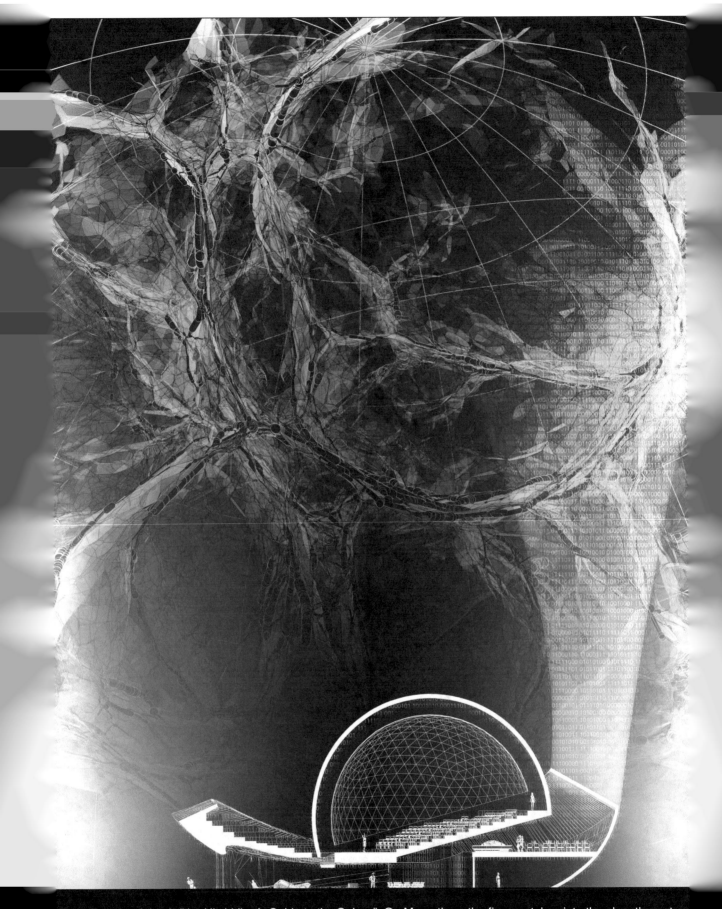

From Douglas Adam's "the Hitchhiker's Guide to the Galaxy": On Magrathea, the five are taken into the planet's centre by a man named Slartibartfast. There, they learn that in the distant past a race of "hyperintelligent, pan-dimensional beings" created a supercomputer named Deep Thought to determine the answer to the "Ultimate Question to Life, the Universe, and Everything", which Deep Thought determined to be the number 42. Deep Thought tells its creators that the answer makes no sense to them because they didn't know what the "Ultimate Question" had been in the first place, and offers to design an even greater computer to determine what the Ultimate Question was. *Credit* Niels Bos

IT FROM BIT FROM 4π FROM TI

4π IN THE SKY

While "The Hitchhiker's guide to the galaxy" refers to the number 42 as the answer to the ultimate questions, in the Information Universe a new answer emerges: **4π** — 4π being the surface area of a sphere with unit radius. In physics, information, say **bits**, is now increasingly perceived as a fundamental building block, and the **holographic principle**[29) 30)] states that the total information content of a *spherical volume* such as a **black hole** is not determined by its volume, but by the amount of bits you can map on its *spherical surface* — 4π. So, one dimension (from 3D to 2D) is gone — indeed the characteristic of a hologram.

4π is one of the fundamental constants of our universe. It actually reflects the flatness of the geometry of our universe, which in turn seems required to produce humans able to read a book — the anthropic principle appears to select from the **multiverse** the 4π geometry of our universe. But, the **anthropic principle** might be a weak and easy argument. Stronger is that in Verlinde's theory of **Emergent Gravity**: **dark matter** and **dark energy** are based on the holographic principle with 4π bits on the surface of a sphere. Also, for **black-body** radiation, the third big breakthrough in the 20th century, I show that the intensity of the radiation can be derived in terms of 4π information units on a surface.

Notably, also in the in vitro Information Universe, the 4π approach in technology seems to match nature in vivo optimally:

nothing tops a good show in a digital planetarium (well, it's 2π) and also the enormous success of the **Virtual Observatory**, connecting 1,000s of databases worldwide, is facilitated by making unique **addresses** on the sky on a 4π surface, using innovative triangular meshes.

But where does this 4π come from? Padmanabhan gives his answer in this book. **CosmIn** carried the information 4π from the pre-existing **Ti** during the **Big Melt** (perceived as the **Big Bang** by many other physicists) into our universe. By CosmIn one can relate to the numerical value of cosmological constant **Λ**. Altogether, the ideas collected in the Information Universe combine into a notion: **"It from bit from 4π from Ti"**.

4π
Bit
Holography
Black hole
Anthropic principle
Multiverse
Emergent Gravity
Dark matter
Dark energy
Black-body radiation
Virtual Observatory
Links — addresses
CosmIn
Ti
Big Melt
It from bit
Bit from Ti
Big Bang

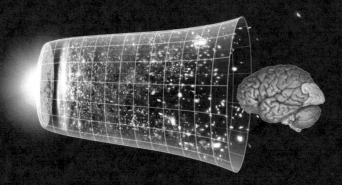

Information only exists in relation to its environment by being copied

ERIK VERLINDE IN CHASING EINSTEIN — THE MOVIE:

"To understand my theory of Emergent Gravity, you need to look at how the language of science has been changing over the centuries. In the 19th century people had mechanical machines, they had steam engines, and thus people had very mechanical ideas also in physics. Then came the 20th century, when people discovered electrons, and it became a century of particles and forces. But now we are in the 21st century, and the language of physics again changes. We live in an information age; so we start thinking about bits and the building blocks of information. And then you start thinking: maybe the Universe is just some kind of giant computer, where information is processed and moved around. Then particles become information carriers. Indeed, I had an idea that space was filled with information, and that *that* was where gravity came from. I started seeing a mental picture of galaxies rotating in this "sea" of information. The way a galaxy looks, if you really think about it, can be compared to a hurricane: the way you see clouds moving in a certain way. You know that it's the air rotating, and that there are a lot of things happening which we don't see: the clouds are just a layer on top. Similarly, I see a galaxy as only that which we can observe; but there is something moving around it. So my picture of a galaxy is not that it resides in some "empty space"; no, it's like a little whirlpool. The Universe is filled with dark energy, and there are small pockets of ordinary matter floating around. A galaxy is where the energy has been localized into a clump of matter. But this has removed part of the energy around it, and that wants to restore itself. And that's what gives the force that we observe as gravity, that people now say is dark matter. But it's actually just due to the dark energy pushing back, trying to fill up the void that is created by the energy that is localized. Then you don't need particle dark matter to explain those observations. So my prediction is that, even when people keep searching for particle dark matter, they won't find anything. And so far, so good; I mean, at some moment when it has not been found again by even better or bigger experiments, people have to start thinking about other possibilities."

Credit Chasing Einstein, Steve Brown & Timothy Wheeler, Ignite Channel, 2019

THE INFORMATION UNIVERSE

A key item in the Information Universe was first recognized by **Shannon**[24) in 1948: information gets a meaning when it is being copied, in other words: *information only exists in relation to its environment by being copied*. Indeed IT — Information Technology — equals ICT — Information Communication Technology. "Information only exists in relation to its environment by being copied" gets a deeper meaning in a **4π** Universe in which information is counted on two-dimensional surfaces and scales with R^{-2} instead of with volume — the **holographic principle**. In fact, the underlying law is that our universe strives to spread information over maximal space by means of an area law, which scales with R^{-2} for spheres, and solid angle for — for instance — **black-body radiation**. It appears that this is a most fundamental property of our Universe: its expansion, its ever increasing amount of entropy and information all seem to be phenomena of spreading information over a maximum of space.

The next key notion is then: *"An object is equal to the information describing it"*. For our computers (in vitro) we know everything works according to the bits and bytes we throw in. Is there a fundamental difference between **in vivo** (in nature) and **in vitro** (machines and observations)? Is 't Hooft right when he asked: "Is the universe one big information processing machine?". In this book I collected a number of approaches by scientists who worked on this theme and came with surprising results: Verlinde's Emergent Gravity, Padmanabhan's **CosmIn** and my own extreme data lineage approach for spreadsheeting downloads of the universe.

Another approach was given in Frieden's early work[10,11)] showing that most physical laws can be derived by minimizing *I* - *J* from **Fisher information,** with

- *I* being the information squeezed out of Nature — what we observe and measure — the phenomenon — Kant: Die Erscheinung, and
- *J* being what nature knows — contained in Nature — intrinsic information — the noumenon — Kant: Das Ding an Sich.

In the end, minimizing *I* - *J* corresponds to minimizing the difference between in vivo – in vitro, and the two worlds of phenomenon and noumenon become indistinguishable. Also Frieden wonders in the end whether noumenon and phenomenon are interchangeable. In this book I collected quite a number of **cross-overs** in vivo — in vitro, and they appear to illustrate that the difference is very thin indeed. Simple but strong examples are the Intihuatana stone in **Machu Picchu**, the incredible train of in vivo – in vitro signal processing in **the ear** and, simply, the story of the lost boy in India, which is just an extreme version of our experiences in daily life.

4π	**The ear**
Dark matter	**Fisher information**
Dark energy	**Shannon**
Holography	**Being copied**
In vivo	**Black-body radiation**
In vitro	**Emergent Gravity**
Cross-over	**CosmIn**
Machu Picchu	

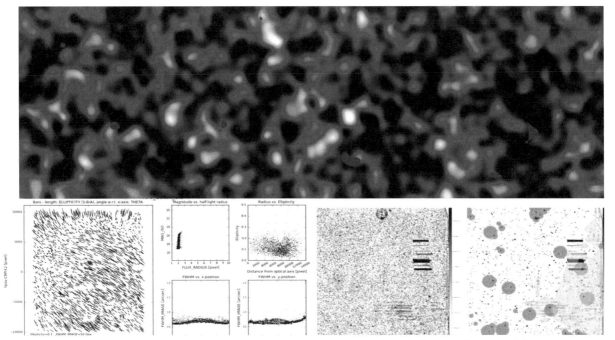

Credit OmegaCAM - KiDS collaboration

ASTRONOMICAL IMAGING

Astronomical imaging surveys like KiDS and the Euclid satellite collect many 100,000s of large digital images, which have to be carefully cleaned for dozens of instrumental effects (footprints), of which several depend on time or position in the field of view, and even on cosmic weather. Many of these effects are slightly different on each image. All the effects are measured, analyzed, discussed, and computer code is designed, specified, implemented and qualified to check, image by image, the zoo of effects. All this happens according to procedures and rules, but also according to the often unwritten rules of the System of Science researchers have been trained to follow. This System of Science is the human-built vehicle to create our facts according to these rules, established by centuries of evolution and given (copied) from generation to generation.

In the era of Big Data this process gets more and more demanding. We need computers and also Artificial Intelligence to perform these tasks, not only for research, but also for any media data that are accumulated in the current Big Data era.

In the last 50 years we went through many hypes in IT: PCs, Grid computing, Big Data and currently Machine learning / Artificial Intelligence. My prediction[38] is that the next hype will be all about **data validation**, a chique word for discriminating Facts and Fakes.

Das *Wahre* und *Falfche* gehört zu den beſtimmten Gedanken, die bewegungslos für eigne Wefen gelten, deren eines drüben, das andre hüben ohne Gemeinfchaft mit dem andern ifolirt und feſt ſteht. Dagegen mufs behauptet werden, dafs die Wahrheit nicht eine ausgeprägte Münze ift, die fertig gegeben, und fo eingeftrichen werden kann. Noch *gibt* es ein Falfches fo wenig es ein Böfes gibt. So fchlimm zwar als der Teufel ift das Böfe und Falfche nicht, denn als diefer find fie fogar zum befondern *Subjecte* gemacht; als Falfches und Bö-

Snippet from Hegel's "Die Phanomenologie des Geistes" (1807) in which he discusses "Das Wahre und Falshe", meaning Facts and Fakes.

System of Science
Facts and Fakes
Data-centric
Open Science
In vivo
Big Data
Artificial Intelligence

FACTS AND FAKES

Science is fluid: our progress in knowledge and insights can be taken as a continuous process, like a flowing river — David Bohm[36]. We believe things to be true up to a point when it is shown not to be true any more. We have devised a **System of Science** to channel the river. Hegel[37] was one of the first to recognize that scientists devise their own system for validation in order to substantiate a reality. And it works marvelously. Remarkably, more than 200 years ago in that same work, Hegel discussed "Das Wahre und Falsche" = "**Facts and Fakes**" and noted that facts are not cast in concrete. Reality, in the end, is cast by consciousness — true is what we believe to be true according to any of those systems, be it a formal, private or home-made system. The consensus of a society might help to settle a truth, but even there we see a variety of truths — between countries and even within countries and politics. Dangerous cultures are run by persons who got into power without any education in how knowledge systems work, or who know it very well and create their own version.

These diversions of reality are now amplified in a world in which information is floating around like crazy on the internet. How can we ever know which System of Science has led to a particular piece of information? This is now referred to as the **Facts and Fakes** problem, mostly on internet. Today, this problem applies to news and social media, but also to the scientific endeavor. The big scientific publishers receive several million submitted articles a year, and the reviewing of these papers by peers has become extremely demanding. I am researching these problems with a team, starting

with the notion that the abstractions and methods we devised for validating scientific Big Data could be applied to the digital media.

There are two fundamentally different approaches:

1) building **data-centric** information systems, in which each data item can be traced back to its sources — *see* page 83 "**The Universe as a spreadsheet**". This is how journalists learn to do their fact checking: check at the source. Building such information systems is extremely demanding, but the result is really what you want. Everything which went into the result can be traced back to the source, which is in fact nature **in vivo** in many cases. This is the ultimate goal of the current European and world wide **Open Science** initiative. It will take many years to have this fully implemented for many different disciplines.

2) For data behind a wall, so NOT open science, be it deliberate (via a VPN or what-ever internet wall) or unreachable in ignorance, we cannot trace the source and we have to devise a completely different technique using **Artificial Intelligence**. We check the heartbeat and all other possible symptoms in a paragraph of text, and determine (like a lie detector) the overall state of a collection of items, resulting in a recommendation on a **Fact and Fake** scale.

ATCGAACACTGAATCGATCTTCAGTATCCAAGTCAGGATTTACGAGGGTATACATATCGGAGG
ATTATTTAATACGTGAGTCCTGATGACATGGATGACAGATCCTAA

LIFE — HOW DOES IT WORK IN THE INFORMATION UNIVERSE?

The key ingredients of life are **DNA** and RNA. Both contain a long string of nucleic acids that come in pairs only, a single strand for RNA and a double helix for DNA. Nevertheless, in both cases the length of the strings is expressed in base pairs. A base pair can have four states: e.g. A-T, T-A, C-G and G-C. A bit has two states: zero or one, so a single base pair corresponds to two bits of information. The amazing thing of the genome is that the base pairs on these strings are glued together to form one unity, like the Intihuatana stone in **Machu Picchu**. This unity of the string acts as an enormous address. While for the **in vitro ASCII** code we use seven bits to address $2^7 = 128$ letters and symbols, nature does this in vivo by using enormously long strings, which result in an incredible and unmeasurable amount of potential addresses — states in physics —

For **COVID-19** with its 30 kilobase pairs, so 60 kilo-bits of information, we obtain $10^{18,072}$ possible states for RNA strings of this length, while the human DNA string, with its 3×10^9 base pairs, which equals 200 bibles, could potentially address $10^{1,807,228,917}$ different states. These are absolutely crazy numbers, nearly none of all these possible states will ever occur, but it demonstrates the extreme complexity of the nature of Life. Since the emergence of **multicellular** organisms about a billion years ago, on average three incremental mutations per year have led to the current DNA (ignoring for simplicity multiple mutations at the same location in the genome) which means that every four years of evolution, the amount of possible states of the current genome was reduced by a factor $8^4 = 4096$, over a period of a billion years.

LIFE

When I started writing this book I expected the last page would be fully devoted to the typical big astronomical numbers like the billions of stars in galaxies, the 100 billion galaxies in our Universe, topped by the number of 10^{77} photons in the **Cosmic Microwave Background (CMB)** and Verlinde's speculative assignment of 10^{120} bits to **dark energy**, and to subsequently sobering this thought by noting that a barcode of only 256 lines has a unique code for every particle in the Universe. In Powers of Two we always describe a theme by its number of possible **states**. Amazingly, the number of possible states in Life is often totally off-scale when compared to the big astronomical numbers: in Life we are dealing with barcodes much longer than 256 lines: up to billions in **DNA**.

The **human brain** with its 10^{11} neurons and 10^{15} synapses hosts a complexity, **in vivo**, which is totally unreachable **in vitro**: no way to get near to even a small fraction of the brain's processing power with machines. Karlheinz Meier, coordinator of the EU Human Brain project, stated: "the transistors in digital computers consume 100,000 billion times more energy than the synapses of neurons". Though the neural networks of the brain are a good guide to designing computers, there is no way Artificial Intelligence (AI) can get anywhere near the power of the human brain. **AI** is good in automating tasks that humans find boring or too demanding. On the other hand, the **complexity** of Life and its computing power could on the long term be paralleled by quantum computing: **qubits** are put in a virtual string like **DNA**, by **entangling** them. When such computers can really be made and can control their own errors, they might trigger a technical revolution indeed.

When we view a full human **DNA** sequence as the barcode of a person, then such a barcode can describe an unmeasurable number of different beings. What does this mean?

While the trick of gluing information in long strings is very smart, these strings only evolve when their replication (copying of information) is imperfect and makes errors. These errors are called mutations. These mutations lead to either a "full stop" to further replications in most cases, or to a "go" in cases when the product fits the environment: first of all to the **4π** of the flat space of our universe, which means, for example, two **eyes** and two **ears** to determine distances, and two or four legs not to fall over on land. The very few surviving cases out of the plethora of possibilities mirror the properties of the local environment. Human DNA must be a footprint of mother Earth. Though the same process can occur on other planets, this notion emphasizes the uniqueness of life on Earth. The fact that for every four years of the past billion years of evolution the amount of possible appearances of human **Life** is reduced by a factor ~4,000 calls into question whether counting planets is a useful technique to assess the probability of intelligent Life in our Universe. Intelligent life might be unique.

For now intelligent Life on Earth, wondering about its origins, appears truly unique in our Universe, perhaps according to the **Anthropic principle** in a **multiverse**.

APPENDIX

POWERS OF TWO IN A TABLE

 # bits = number of bits in a single string = x

*# states = number of unique addresses (or states) that can be identified (made)
with that string = 2ˣ = Powers of Two*

 # bytes for which this string can make unique addresses

	# bits	#states	#bytes	bit	byte	
	0	1.0e+00	0.0e+00			Multiverse
bit	1	2.0e+00	0.0e+00			Big Bang
	2	4.0e+00	0.0e+00			Mullticellular
	3	8.0e+00	1.0e+00			
	4	1.6e+01	2.0e+00			
	5	3.2e+01	4.0e+00			
	6	6.4e+01	8.0e+00			
alphabeth	7	1.3e+02	1.6e+01			ASCII
byte	8	2.6e+02	3.2e+01			Machu Picchu
	9	5.1e+02	6.4e+01			
	10	1.0e+03	1.3e+02			
	11	2.0e+03	2.6e+02			
	12	4.1e+03	5.1e+02			
	13	8.2e+03	1.0e+03			
	14	1.6e+04	2.0e+03			
	15	3.3e+04	4.1e+03	KILO		
2 bytes	16	6.6e+04	8.2e+03			Covid-19/first cpu
	17	1.3e+05	1.6e+04			
	18	2.6e+05	3.3e+04			
	19	5.2e+05	6.6e+04			
	20	1.0e+06	1.3e+05			
	21	2.1e+06	2.6e+05			
	22	4.2e+06	5.2e+05			
	23	8.4e+06	1.0e+06			
	24	1.7e+07	2.1e+06	MEGA		hard disks
	25	3.4e+07	4.2e+06			
	26	6.7e+07	8.4e+06			
	27	1.3e+08	1.7e+07			
	28	2.7e+08	3.4e+07			
	29	5.4e+08	6.7e+07			
	30	1.1e+09	1.3e+08			
	31	2.1e+09	2.7e+08			
4 bytes	32	4.3e+09	5.4e+08			human DNA
	33	8.6e+09	1.1e+09			DVD
	34	1.7e+10	2.1e+09	GIGA		
	35	3.4e+10	4.3e+09			bio banking
	36	6.9e+10	8.6e+09			human brain
	37	1.4e+11	1.7e+10			
	38	2.7e+11	3.4e+10			GAIA/CT Scan
	39	5.5e+11	6.9e+10			
	40	1.1e+12	1.4e+11			

	# bits	#states	#bytes	bit	byte	
	41	2.2e+12	2.7e+11			
	42	4.4e+12	5.5e+11			
	43	8.8e+12	1.1e+12			Terabyte
	44	1.8e+13	2.2e+12			
	45	3.5e+13	4.4e+12	TERA		
	46	7.0e+13	8.8e+12			
	47	1.4e+14	1.8e+13			
	48	2.8e+14	3.5e+13			
	49	5.6e+14	7.0e+13			
	50	1.1e+15	1.4e+14			
	51	2.3e+15	2.8e+14			
	52	4.5e+15	5.6e+14			
	53	9.0e+15	1.1e+15			Petabyte
	54	1.8e+16	2.3e+15			
	55	3.6e+16	4.5e+15	PETA		
	56	7.2e+16	9.0e+15			
	57	1.4e+17	1.8e+16			
	58	2.9e+17	3.6e+16			
	59	5.8e+17	7.2e+16			
	60	1.2e+18	1.4e+17			
	61	2.3e+18	2.9e+17			
	62	4.6e+18	5.8e+17			100's Petabyte
	63	9.2e+18	1.2e+18			
8 bytes	64	1.8e+19	2.3e+18			Exabyte - SKA
	65	3.7e+19	4.6e+18			
	66	7.4e+19	9.2e+18	EXA		CERN 2020s
	67	1.5e+20	1.8e+19			
	68	3.0e+20	3.7e+19			
	69	5.9e+20	7.4e+19			
	70	1.2e+21	1.5e+20			
	71	2.4e+21	3.0e+20			
	72	4.7e+21	5.9e+20			
	73	9.4e+21	1.2e+21			
	74	1.9e+22	2.4e+21			
	75	3.8e+22	4.7e+21			
	76	7.6e+22	9.4e+21			
	77	1.5e+23	1.9e+22			
	78	3.0e+23	3.8e+22			
	79	6.0e+23	7.6e+22			
	80	1.2e+24	1.5e+23			

# bits	#states	#bytes
81	2.4e+24	3.0e+23
82	4.8e+24	6.0e+23
83	9.7e+24	1.2e+24
84	1.9e+25	2.4e+24
85	3.9e+25	4.8e+24
86	7.7e+25	9.7e+24
87	1.5e+26	1.9e+25
88	3.1e+26	3.9e+25
89	6.2e+26	7.7e+25
90	1.2e+27	1.5e+26
91	2.5e+27	3.1e+26
92	5.0e+27	6.2e+26
93	9.9e+27	1.2e+27
94	2.0e+28	2.5e+27
95	4.0e+28	5.0e+27
96	7.9e+28	9.9e+27
97	1.6e+29	2.0e+28
98	3.2e+29	4.0e+28
99	6.3e+29	7.9e+28
100	1.3e+30	1.6e+29
101	2.5e+30	3.2e+29
102	5.1e+30	6.3e+29
103	1.0e+31	1.3e+30
104	2.0e+31	2.5e+30
105	4.1e+31	5.1e+30
106	8.1e+31	1.0e+31
107	1.6e+32	2.0e+31
108	3.2e+32	4.1e+31
109	6.5e+32	8.1e+31
110	1.3e+33	1.6e+32
111	2.6e+33	3.2e+32
112	5.2e+33	6.5e+32
113	1.0e+34	1.3e+33
114	2.1e+34	2.6e+33
115	4.2e+34	5.2e+33
116	8.3e+34	1.0e+34
117	1.7e+35	2.1e+34
118	3.3e+35	4.2e+34
119	6.6e+35	8.3e+34
120	1.3e+36	1.7e+35

	# bits	#states	#bytes		
	121	2.7e+36	3.3e+35		
	122	5.3e+36	6.6e+35		
	123	1.1e+37	1.3e+36		
	124	2.1e+37	2.7e+36		
	125	4.3e+37	5.3e+36		
	126	8.5e+37	1.1e+37		
	127	1.7e+38	2.1e+37		
16 bytes	128	3.4e+38	4.3e+37		Cryptography
	129	6.8e+38	8.5e+37		
	130	1.4e+39	1.7e+38		
	131	2.7e+39	3.4e+38		
	132	5.4e+39	6.8e+38		
	133	1.1e+40	1.4e+39		
	134	2.2e+40	2.7e+39		
	135	4.4e+40	5.4e+39		
	136	8.7e+40	1.1e+40		
	137	1.7e+41	2.2e+40		
	138	3.5e+41	4.4e+40		
	139	7.0e+41	8.7e+40		
	140	1.4e+42	1.7e+41		
	141	2.8e+42	3.5e+41		
	142	5.6e+42	7.0e+41		
	143	1.1e+43	1.4e+42		
	144	2.2e+43	2.8e+42		
	145	4.5e+43	5.6e+42		
	146	8.9e+43	1.1e+43		
	147	1.8e+44	2.2e+43		
	148	3.6e+44	4.5e+43		
	149	7.1e+44	8.9e+43	the	
	150	1.4e+45	1.8e+44	DESERT	
	151	2.9e+45	3.6e+44		
	152	5.7e+45	7.1e+44		
	153	1.1e+46	1.4e+45		
	154	2.3e+46	2.9e+45		
	155	4.6e+46	5.7e+45		
	156	9.1e+46	1.1e+46		
	157	1.8e+47	2.3e+46		
	158	3.7e+47	4.6e+46		
	159	7.3e+47	9.1e+46		
	160	1.5e+48	1.8e+47		

# bits	#states	#bytes			
161	2.9e+48	3.7e+47			
162	5.8e+48	7.3e+47			
163	1.2e+49	1.5e+48			
164	2.3e+49	2.9e+48			
165	4.7e+49	5.8e+48			
166	9.4e+49	1.2e+49			
167	1.9e+50	2.3e+49			
168	3.7e+50	4.7e+49			
169	7.5e+50	9.4e+49			
170	1.5e+51	1.9e+50			
171	3.0e+51	3.7e+50			
172	6.0e+51	7.5e+50			
173	1.2e+52	1.5e+51			
174	2.4e+52	3.0e+51			
175	4.8e+52	6.0e+51			
176	9.6e+52	1.2e+52			
177	1.9e+53	2.4e+52			
178	3.8e+53	4.8e+52			
179	7.7e+53	9.6e+52			
180	1.5e+54	1.9e+53	the		
181	3.1e+54	3.8e+53	DESERT		
182	6.1e+54	7.7e+53			
183	1.2e+55	1.5e+54			
184	2.5e+55	3.1e+54			
185	4.9e+55	6.1e+54			
186	9.8e+55	1.2e+55			
187	2.0e+56	2.5e+55			
188	3.9e+56	4.9e+55			
189	7.8e+56	9.8e+55			
190	1.6e+57	2.0e+56			
191	3.1e+57	3.9e+56			
192	6.3e+57	7.8e+56			
193	1.3e+58	1.6e+57			
194	2.5e+58	3.1e+57			
195	5.0e+58	6.3e+57			
196	1.0e+59	1.3e+58			
197	2.0e+59	2.5e+58			
198	4.0e+59	5.0e+58			
199	8.0e+59	1.0e+59			
200	1.6e+60	2.0e+59			

# bits	#states	#bytes			
201	3.2e+60	4.0e+59			
202	6.4e+60	8.0e+59			
203	1.3e+61	1.6e+60			
204	2.6e+61	3.2e+60			
205	5.1e+61	6.4e+60			
206	1.0e+62	1.3e+61			
207	2.1e+62	2.6e+61			
208	4.1e+62	5.1e+61			
209	8.2e+62	1.0e+62			
210	1.6e+63	2.1e+62			
211	3.3e+63	4.1e+62			
212	6.6e+63	8.2e+62			
213	1.3e+64	1.6e+63			
214	2.6e+64	3.3e+63			
215	5.3e+64	6.6e+63			
216	1.1e+65	1.3e+64			
217	2.1e+65	2.6e+64			
218	4.2e+65	5.3e+64			
219	8.4e+65	1.1e+65			
220	1.7e+66	2.1e+65	the		
221	3.4e+66	4.2e+65	DESERT		
222	6.7e+66	8.4e+65			
223	1.3e+67	1.7e+66			
224	2.7e+67	3.4e+66			
225	5.4e+67	6.7e+66			
226	1.1e+68	1.3e+67			
227	2.2e+68	2.7e+67			
228	4.3e+68	5.4e+67			
229	8.6e+68	1.1e+68			
230	1.7e+69	2.2e+68			
231	3.5e+69	4.3e+68			
232	6.9e+69	8.6e+68			
233	1.4e+70	1.7e+69			
234	2.8e+70	3.5e+69			
235	5.5e+70	6.9e+69			
236	1.1e+71	1.4e+70			
237	2.2e+71	2.8e+70			
238	4.4e+71	5.5e+70			
239	8.8e+71	1.1e+71			
240	1.8e+72	2.2e+71			

# bits	#states	#bytes			
241	3.5e+72	4.4e+71			
242	7.1e+72	8.8e+71			
243	1.4e+73	1.8e+72			
244	2.8e+73	3.5e+72			
245	5.7e+73	7.1e+72			
246	1.1e+74	1.4e+73			
247	2.3e+74	2.8e+73			
248	4.5e+74	5.7e+73			
249	9.0e+74	1.1e+74			
250	1.8e+75	2.3e+74			
251	3.6e+75	4.5e+74			
252	7.2e+75	9.0e+74			
253	1.4e+76	1.8e+75			
254	2.9e+76	3.6e+75			
255	5.8e+76	7.2e+75			
32 bytes	256	1.2e+77	1.4e+76		CMB/everything
	257	2.3e+77	2.9e+76		
	258	4.6e+77	5.8e+76		black hole
	259	9.3e+77	1.2e+77		
	260	1.9e+78	2.3e+77		
	261	3.7e+78	4.6e+77		
	262	7.4e+78	9.3e+77		
	263	1.5e+79	1.9e+78		
	264	3.0e+79	3.7e+78		
	265	5.9e+79	7.4e+78		
	266	1.2e+80	1.5e+79		
	267	2.4e+80	3.0e+79		
	268	4.7e+80	5.9e+79		
	269	9.5e+80	1.2e+80		
	270	1.9e+81	2.4e+80		
	271	3.8e+81	4.7e+80		
	272	7.6e+81	9.5e+80		
	273	1.5e+82	1.9e+81		
	274	3.0e+82	3.8e+81		
	275	6.1e+82	7.6e+81		
	276	1.2e+83	1.5e+82		
	277	2.4e+83	3.0e+82		
	278	4.9e+83	6.1e+82		
	279	9.7e+83	1.2e+83		
	280	1.9e+84	2.4e+83		

# bits	#states	#bytes		
281	3.9e+84	4.9e+83		
282	7.8e+84	9.7e+83		
283	1.6e+85	1.9e+84		
284	3.1e+85	3.9e+84		
285	6.2e+85	7.8e+84		
286	1.2e+86	1.6e+85		
287	2.5e+86	3.1e+85		
288	5.0e+86	6.2e+85		
289	9.9e+86	1.2e+86		
290	2.0e+87	2.5e+86		
291	4.0e+87	5.0e+86		
292	8.0e+87	9.9e+86		
293	1.6e+88	2.0e+87		
294	3.2e+88	4.0e+87		
295	6.4e+88	8.0e+87		
296	1.3e+89	1.6e+88		
297	2.5e+89	3.2e+88		
298	5.1e+89	6.4e+88		
299	1.0e+90	1.3e+89		
300	2.0e+90	2.5e+89		
301	4.1e+90	5.1e+89		
302	8.1e+90	1.0e+90		
303	1.6e+91	2.0e+90		
304	3.3e+91	4.1e+90		
305	6.5e+91	8.1e+90		
306	1.3e+92	1.6e+91		
307	2.6e+92	3.3e+91		
308	5.2e+92	6.5e+91		
309	1.0e+93	1.3e+92		
310	2.1e+93	2.6e+92		
311	4.2e+93	5.2e+92		
312	8.3e+93	1.0e+93		
313	1.7e+94	2.1e+93		
314	3.3e+94	4.2e+93		
315	6.7e+94	8.3e+93		
316	1.3e+95	1.7e+94		
317	2.7e+95	3.3e+94		
318	5.3e+95	6.7e+94		
319	1.1e+96	1.3e+95		
320	2.1e+96	2.7e+95		

# bits	#states	#bytes
321	4.3e+96	5.3e+95
322	8.5e+96	1.1e+96
323	1.7e+97	2.1e+96
324	3.4e+97	4.3e+96
325	6.8e+97	8.5e+96
326	1.4e+98	1.7e+97
327	2.7e+98	3.4e+97
328	5.5e+98	6.8e+97
329	1.1e+99	1.4e+98
330	2.2e+99	2.7e+98
331	4.4e+99	5.5e+98
332	8.7e+99	1.1e+99
333	1.7e+100	2.2e+99
334	3.5e+100	4.4e+99
335	7.0e+100	8.7e+99
336	1.4e+101	1.7e+100
337	2.8e+101	3.5e+100
338	5.6e+101	7.0e+100
339	1.1e+102	1.4e+101
340	2.2e+102	2.8e+101
341	4.5e+102	5.6e+101
342	9.0e+102	1.1e+102
343	1.8e+103	2.2e+102
344	3.6e+103	4.5e+102
345	7.2e+103	9.0e+102
346	1.4e+104	1.8e+103
347	2.9e+104	3.6e+103
348	5.7e+104	7.2e+103
349	1.1e+105	1.4e+104
350	2.3e+105	2.9e+104
351	4.6e+105	5.7e+104
352	9.2e+105	1.1e+105
353	1.8e+106	2.3e+105
354	3.7e+106	4.6e+105
355	7.3e+106	9.2e+105
356	1.5e+107	1.8e+106
357	2.9e+107	3.7e+106
358	5.9e+107	7.3e+106
359	1.2e+108	1.5e+107
360	2.3e+108	2.9e+107

# bits	#states	#bytes		
362	9.4e+108	1.2e+108		
363	1.9e+109	2.3e+108		
364	3.8e+109	4.7e+108		
365	7.5e+109	9.4e+108		
366	1.5e+110	1.9e+109		
367	3.0e+110	3.8e+109		
368	6.0e+110	7.5e+109		
369	1.2e+111	1.5e+110		
370	2.4e+111	3.0e+110		
371	4.8e+111	6.0e+110		
372	9.6e+111	1.2e+111		
373	1.9e+112	2.4e+111		
374	3.8e+112	4.8e+111		
375	7.7e+112	9.6e+111		
376	1.5e+113	1.9e+112		
377	3.1e+113	3.8e+112		
378	6.2e+113	7.7e+112		
379	1.2e+114	1.5e+113		
380	2.5e+114	3.1e+113		
381	4.9e+114	6.2e+113		
382	9.9e+114	1.2e+114		
383	2.0e+115	2.5e+114		
384	3.9e+115	4.9e+114		
385	7.9e+115	9.9e+114		
386	1.6e+116	2.0e+115		
387	3.2e+116	3.9e+115		
388	6.3e+116	7.9e+115		
389	1.3e+117	1.6e+116		
390	2.5e+117	3.2e+116		
391	5.0e+117	6.3e+116		
392	1.0e+118	1.3e+117		
393	2.0e+118	2.5e+117		
394	4.0e+118	5.0e+117		
395	8.1e+118	1.0e+118		
396	1.6e+119	2.0e+118		
397	3.2e+119	4.0e+118		
398	6.5e+119	8.1e+118		
399	1.3e+120	1.6e+119		Cosmic Event Horizon/ Dark Energy

REFERENCES

1) Linde A D (1986) *Eternally existing self-reproducing chaotic inflationary universe.* Physics Letters B, Vol 175, Iss 4. doi:10.1016/0370-2693(86)90611-8

2) Susskind L (2006) *The Cosmic Landscape: String Theory and the Illusion of Intelligent Design.* Back Bay Books.

3) Pennisi E (2018) *The momentous transition to multicellular life may not have been so hard after all.* Science Mag, AAAS. www.sciencemag.org/news/2018/06/momentous-transition-multicellular-life-may-not-have-been-so-hard-after-all

4) Gleick J (1998) *Chaos: making a new science.* Vintage, London

5) Anthony SJ, Epstein JH, Murray KA, et al. (2013) *A Strategy To Estimate Unknown Viral Diversity in Mammals.* mBio 4:e00598-13. doi:10.1128/mBio.00598-13

6) Lauberts A (1982) *The ESO/Uppsala survey of the ESO (B) atlas.* European Southern Observatory, München

7) Lauberts A, Valentijn EA (1989) *The Surface Photometry Catalogue of the ESO-Uppsala galaxies.* European Southern Observatory, Garching bei München

8) Toward an International Virtual Observatory - Scientific Motivation, Roadmap for Development and Current Status, Garching bei München, Germany, June 10-14, 2002.

9) Axer M, Strohmer S, Gräßel D, Bücker O, Dohmen M, Reckfort J, Zilles K, Amunts K (2016) *Estimating Fiber Orientation Distribution Functions in 3D-Polarized Light Imaging.* 9 Frontiers in Neuroanatomy 10: 1-12. doi:10.3389/fnana.2016.00040

10) Frieden BR (1998) *Physics from Fisher information: a unification.* Cambridge University Press, Cambridge, U.K.; New York, U.S.A.

11) Frieden BR (2004) *Science from Fisher Information: a unification.* Cambridge University Press, Cambridge, U.K.; New York, U.S.A.

12) Springel V, Frenk CS, White SDM (2006) *The large-scale structure of the Universe.* Nature 440:1137–1144. doi:10.1038/nature04805

13) Aladin Sky Atlas. aladin.u-strasbg.fr/

14) Google Sky. www.google.com/sky/

15) AAS WorldWide Telescope. www.worldwidetelescope.org/

16) Humphreys PC, Kalb N, Morits JPJ et al. (2018) *Deterministic delivery of remote entanglement on a quantum network.* Nature 558, 268–273. doi:10.1038/s41586-018-0200-5

17) Boschi D, Branca S, De Martini F, Hardy L, Popescu S (1998). *Experimental Realization of Teleporting an Unknown Pure Quantum State via Dual Classical and Einstein-Podolsky-Rosen Channels. Physical Review Letters. 80 (6): 1121–1125.* doi:10.1103/PhysRevLett.80.1121

18) Einstein A, Podolsky B, Rosen N (1935) *Can Quantum-Mechanical Description of Physical Reality Be Considered Complete?* Phys Rev 47:777–780 . doi:10.1103/PhysRev.47.777

19) Bell JS (1964) *On the Einstein Podolsky Rosen paradox.* Physics Physique Fizika 1:195–200. doi:10.1103/PhysicsPhysiqueFizika.1.195

20) Aspect A (1999) *Bell's inequality test: more ideal than ever*. Nature 398:189–190 . doi:10.1038/18296

21) Shor PW (1994) *Algorithms for quantum computation: discrete logarithms and factoring*. Proceedings 35th Annual Symposium on Foundations of Computer Science. IEEE Comput. Soc. Press: 124–134. 199421)

22) Diffie W, Hellman ME (1976) *New Directions in Cryptography*. IEEE Transactions on Information Theory. 22 (6): 644–654.

23) Merkle, RC (1977) *Secure Communications Over Insecure Channels*. Communications of the ACM. 21 (4): 294–299.

24) Shannon CE (1948) *A Mathematical Theory of Communication*. Bell System Technical Journal 27:379–423. doi:10.1002/j.1538-7305.1948.tb01338.x

25) The Nobel Prize in Physiology or Medicine 1973. www.nobelprize.org/prizes/medicine/1973/frisch/facts/

26) Schurch R, Ratnieks FLW (2015) *The spatial information content of the honey bee waggle dance*. Frontiers in Human Neuroscience 3. doi:10.3389/fevo.2015.00022

27) Sloot MA, Kampis G, Gulyás L (2013) *Advances in dynamic temporal networks: Understanding the temporal dynamics of complex adaptive networks*. The European Physical Journal Special Topics, Volume 222, Issue 6, 2013, pp.1287-1293. doi:10.1140/epjst/e2013-01926-8

28) OmegaCAM. www.eso.org/public/teles-instr/paranal-observatory/surveytelescopes/vst/camera/

29) Bekenstein JD (1973) *Black Holes and Entropy*. Phys Rev D 7, 2333. doi:10.1103/PhysRevD.7.2333.

30) Hawking SW (1975) *Particle creation by black holes*. Communications in Mathematical Physics 43 (3): 199–220. doi:10.1007/BF02345020

31) Herculano-Houzel S (2012) *The remarkable, yet not extraordinary, human brain as a scaled-up primate brain and its associated cost*. Proc Natl Acad Sci USA, 109 (Suppl 1): 10661–10668. doi:10.1073/pnas.1201895109

32) Jaynes ET (1957) *Information Theory and Statistical Mechanics*. Phys Rev 106:620–630. doi:10.1103/PhysRev.106.620

33) Frieden BR and Gatenby RG (2020) arXiv:1909.11435 (physics)

34) Gottardi S (2020) in preparation

35) Tegmark M (2017) *Life 3.0* Penguin Random House

36) Bohm D (1980) *Wholeness and the implicate order*, Routledge

37) Hegel GWF (1807) *System der Wissenschaft, Die Phänomenologie des Geistes*, Vorrede page XLV, Bamberg and Würzburg: Goebhardt.

38) Valentijn EA (2018) *Data validation beyond Big Data*, VST in the Era of the Large Sky Surveys, Proceedings of the conference held 5–8 June, 2018 in Naples, Italy, doi:10.5281/zenodo.1303323

39) Crumpler. https://www.csis.org/blogs/technology-policy-blog/how-accurate-are-facial-recognition-systems-%E2%80%93-and-why-does-it-matter

ILLUSTRATIONS

ALL ILLUSTRATIONS BY
EDWIN A. VALENTIJN
EXCEPT THE FOLLOWING ONES:

Chapter	Credit Line
What is Information?	Jacques Descloitres, MODIS Rapid Response Team, NASA/GSFC
Ti — Spacetime foam	NASA/WMAP Science Team
Multiverse — Anthropic principle	6222336 Algol | Dreamstime.com
Big bang	Right page NASA/WMAP Science Team
What is a bit?	Ingram Publishing
Multicellular Life — In vivo	2020 Flinn scientific
	Hisayoshi Nozaki, Yoko Arakaki
The Game of Life — In vitro	Copyright 1996-2004 Edwin Martin. The Game of Life is invented by John Conway
	Courtesy Diana Conway
First computers	Chipdb.org
	The Rise of Radio Astronomy in the Netherlands: The People and the Politics, by Astrid Elbers
	Rama & Musée Bolo (https://commons.wikimedia.org/wiki/File:IBM_PC-IMG_7271.jpg)
COVID-19	Graphen
	Graphen
Star peace - Machu Picchu at the Canaries	Royal Greenwich Obervatory
Pre-internet - Facts and Fakes	NFP
	Queen Beatrix, David Calvert
	Datagraver.com 2020, Data: START GTD
	Royal Greenwich Obervatory, David Calvert
Hard disk	Courtesy International Business Machines Corporation (IBM)
	Courtesy International Business Machines Corporation (IBM)
The telephone — In vivo – in vitro	By the U.S. National Archives
DNA	ANP, Prof Oscar Miller/Science PhotoLibrary
	Gwen Childs
	DNA Strands via Wikimedia Commons (commons.wikimedia.org/wiki/File:DNA_strands.png)
From 1D DNA to 3D proteins	Sten André via Wikimedia Commons (commons.wikimedia.org/wiki/Category:Genetic_code#/media/File:RNA-kodon.png)
	Boumphreyfr via Wikimedia Commons
	https://commons.wikimedia.org/wiki/File:Peptide_syn.png
	Reproduced from Open Clip Art
Where does biological information come from?	Charles Lineweaver
	Charles Lineweaver
	Carl Spitzweg
DVD	CHRONOZOOM
	ESO
	DVD-multimedia / Philips
Biobanking for human health	Joeri van der Velde et al., medRxiv 19012229;
	Joeri van der Velde et al., https://onlinelibrary.wiley.com/doi/abs/10.1002/ggn2.10023
	Joeri van der Velde with embedded credits to Noun project.
	LifeLines B.V.
Human Brain	All Heidelberg University
Neuromorphic computing	All Heidelberg University
Dark Energy — Dark Matter	Alessandra Silvestri, adapted from Marco Raveri et al. (2016)
Gaia — A billion stars	Reyer Boxem, Dagblad van het Noorden
	ESA
	ESA
CT scan	Image courtesy of Sectra AB
	Data Courtesy: Center for Medical Image Science and Visualization (CMIV)
Terabyte	Top right: G. Sikkema for the OmegaCAM consortium
	bottom right VST mirror: ESO
	KiDS collaboration
Paranal Observatory	ESO
Gravitational Lensing	Andrew Fruchter (STScI) et al., WFPC2, HST, NASA
From Terabytes to two numbers	Alex Tudorica, KiloDegree Survey
	APS/Alan Stonebraker; galaxy images from STScI/AURA, NASA, ESA, and the Hubble Heritage Team
	Margot Brouwer et al. 2017 (https://arxiv.org/pdf/1612.03034.pdf)
	Margot Brouwer: Screenshot of KiDS data table
The Euclid satellite	NASA WMAP Science Team
	NASA, ESA, and G. Illingworth and D. Magee (University of California, Santa Cruz)
	NASA/GSFC/Arizona State University
	ESA
Big Data	Images from the diaries of Leo Polak (1901), from the University Library Amsterdam, (NL-AsdUvA_UBAinv373), provided to the Monk system at the University of Groningen (RUG) by Dr Stefan van der Poel
	LOFAR: Netherlands Institute for Radio Astronomy (ASTRON)

CHAPTER	CREDIT LINE
Watching a black hole — In vitro	Katie Bouman, Caltech
	Hotaka Shiokawa, Event Horizon Telescope collaboration
	Event Horizon Telescope collaboration
Watching a black hole — In vivo	Event Horizon Telescope collaboration
Dark Matter maps — In vivo – in vitro	Kids collaboration
Hundreds of petabytes	Paul Butler
	Paul Butler implemented by Jason Sundram
Future - more than hundreds of petabytes	European Space Agency (ESA)
	GIANLUCA LOMBARDI/ LSST/AURA/NSF
The information Universe — In vivo – in vitro	Volker Springer, NATURE (2dFGRS, SDSS, Millenium Simulation/MPA Garching,
	and Gerard Lemson & the Virgo Consortium)
In vivo – in vitro — Story of the lost boy	Google maps:Imagery ©2020 CNES/Airbus. Maxar Technologies. Map data©2020
	Saroo Brierley
Virtual Observatory — Aladin	Centre Donnee Astronomique Strasbourg
Virtual Observatory — IVOA	IVOA
Qubits	QUTech, TU Delft
Quantum computers get real	Klapstuk for QUTech, TU Delft
Entanglement	Doug Cohen and Louis Tarantino from 'Quantum Entanglement Simplified'
	NASA, ESA, H.E. Bond (STScI) and The Hubble Heritage Team (STScI/AURA)
Exabyte — SKA	Both: ICRAR/Curtin
Cryptography	Both: Simona Samardjiska (the first one with added elements from pixabay)
Nature processes Infomation	Emmanuel Boutet / Courtesy: https://en.wikipedia.org/wiki/Bee
The Desert	CERN
	Gerard 't Hooft, modified by Edwin Valentijn
	CERN [https://home.cern/resources/360-image/accelerators/virtual-tour-lhc]
Black Hole	Warren Johnston
Unraveling black holes with gravitational waves	NASA/Goddard/UMBC/Bernard J. Kelly, NASA/Ames/Chris Henze, CSC Government
	Solutions LLC/Tim Sandstrom
	Ron de Wit
The Cosmic Microwave Background	Creative Commons Zero (www.piqsels.com/en/public-domain-photo-fsbro)
	Olcay Ertem. [https://pixabay.com/photos/old-woman-grandmother-portrait-4640282/]
	ESA
	M. Asgari et al. and the KiDS collaboration, 2020, arXiv:2007.15633
Cosmic event horizon	Right page: Pablo Carlos Budassi
Emergent Gravity	F. Pastawski, B. Yoshida, D. Harlow en J. Preskill.
	Angela Miller
Science from Fisher Information	B. Roy Frieden
Multiverse from Fisher Information	Take 27 Ltd
Origins of physical information	Stefano Gottardi, based on andromeda galaxy image from Adam Evans under
	Creative Commons license. https://commons.m.wikimedia.org/wiki/File:
	Andromeda_Galaxy_(with_h-alpha).jpg#mw-jump-to-license
Black-body radiation	Upper left: Erik Verlinde
Cosmic Information, cosmogenesis and	
dark energy	Thanu Padmanabhan
CosmIn and the cosmological constant	Hamsa Padmanabhan, background ESO
System of Science	Kai Froeb modified by Edwin Valentijn
	Mitchell Nolte
The ear — Copying information	SMC468 Graphic Design for Education
Visualization — From in vivo to consciousness	Karin van der Wiel, KNMI
	Ed Hawkins (University of Reading), Attribution 4.0 International (CC BY 4.0)
	https://showyourstripes
Consciousness	Johan Swanepoel
Intelligence	Ton Sijbrands, ANP
Somnium Live	Full spread: Bart Heemskerk
It from bit	Niels Bos
The Information Universe	NASA WMAP Science Team modified by Edwin Valentijn
	CANON
	Chasing Einstein, Steve Brown & Timothy Wheeler, Ignite Channel, 2019.
Facts and Fakes	OmegaCAM - KiDs collaboration
Life	Jan Passchier (https://abstrusegoose.com/275)

Printed in the United States
by Baker & Taylor Publisher Services